This text discusses the internal structure and the e[...] emphasises the basic physics governing stellar stru[...] ideas on which our understanding of stellar structu[...] book also provides a comprehensive discussion of stellar evolution. Careful comparison is made between theory and observation, and the author has thus provided a lucid and balanced introductory text for the student.

Volume 1 in this series by Erika Böhm-Vitense: *Basic stellar observations and data*, introduces the basic elements of fundamental astronomy and astrophysics. Volume 2: *Stellar atmospheres*, conveys the physical ideas and laws used in the study of the outer layers of a star. The present volume is the final one in this short series of books which together provide a modern, complete and authoritative account of our present knowledge of the stars.

Each of the three books is self-contained and can be used as an independent textbook. The author is Professor of Astronomy at the University of Washington, Seattle, and has not only taught but has also published many original papers in this subject. Her clear and readable style should make all three books a first choice for undergraduate and beginning graduate students taking courses in astronomy and particularly in stellar astrophysics.

Introduction to stellar astrophysics

Volume 3
Stellar structure and evolution

Introduction to stellar astrophysics

Volume 3
Stellar structure and evolution

Erika Böhm-Vitense
University of Washington

The right of the
University of Cambridge
to print and sell
all manner of books
was granted by
Henry VIII in 1534.
The University has printed
and published continuously
since 1584.

CAMBRIDGE UNIVERSITY PRESS

Cambridge

New York   Port Chester   Melbourne   Sydney

Published by the Press Syndicate of the University of Cambridge
The Pitt Building, Trumpington Street, Cambridge CB2 1RP
40 West 20th Street, New York, NY 10011-4211, USA
10 Stamford Road, Oakleigh, Victoria 3166, Australia

© Cambridge University Press 1992

First published 1992

Printed in Great Britain at the University Press, Cambridge

British Library cataloguing in publication data

Böhm-Vitense, Erika *1923–*
    Introduction to stellar astrophysics.
    Vol. 3: Stellar structure and evolution
    1. Stars
    I. Title
523.8

Library of Congress cataloguing in publication data

Böhm-Vitense, E.
    Introduction to stellar astrophysics.
    Includes bibliographical references and indexes.
    Contents: v. 1. Basic stellar observations and data – v. 2. Stellar
atmospheres – v. 3. Stellar structure and evolution.
    1. Stars. 2. Astrophysics. I. Title.
QB801.B64 1989   523.8   88–20310

ISBN 0 521 34404 2 hardback
ISBN 0 521 34871 4 paperback

PN

Contents

Preface

In Volume 3 of *Introduction to Stellar Astrophysics* we will discuss the internal structure and the evolution of stars.

Many astronomers feel that stellar structure and evolution is now completely understood and that further studies will not contribute essential knowledge. It is felt that much more is to be gained by the study of extragalactic objects, particularly the study of cosmology. So why write this series of textbooks on *stellar* astrophysics?

We would like to emphasize that 97 per cent of the luminous matter in our Galaxy and in most other galaxies is in stars. Unless we understand thoroughly the light emission of the stars, as well as their evolution and their contribution to the chemical evolution of the galaxies, we cannot correctly interpret the light we receive from external galaxies. Without this knowledge our cosmological derivations will be without a solid foundation and might well be wrong. The ages currently derived for globular clusters are larger than the age of the universe derived from cosmological expansion. Which is wrong, the Hubble constant or the ages of the globular clusters? We only want to point out that there are still open problems which might well indicate that we are still missing some important physical processes in our stellar evolution theory. It is important to emphasize these problems so that we keep thinking about them instead of ignoring them. We might waste a lot of effort and money if we build a cosmological structure on uncertain foundations.

The light we receive from external galaxies has contributions from stars of all ages and masses and possibly very different chemical abundances. We need to study and understand all these different kinds of stars if we want to understand external galaxies.

We feel that we need a short and comprehensive summary of our knowledge about stellar evolution. Much new insight has been gained in the past few decades. In addition, we want to point out that there are still many interesting open questions in the field of stellar structure and evolution: for example, we still do not know accurately what the masses of the Cepheids are. How can we be sure that the period luminosity relation is the same for Cepheids in other galaxies as we observe it in our neighborhood if we are not sure yet that we understand their structure? How can we use supernovae in other galaxies as distance indicators, if we do not understand the dependence of their brightness on mass or the original chemical abundances of the progenitors? Unless we understand the evolution of the presupernovae and the processes which lead to the explosions, we cannot be sure about the intrinsic brightnesses of the supernovae.

Much interesting physics is still to be learned from studying the internal structure of stars. Nowhere in the laboratory can we study such high density matter as in white dwarfs or neutron stars.

In many parts of this volume we shall follow the excellent discussions in the book by M. Schwarzschild (1958) on stellar structure and evolution. Other good books on the topic of stellar structure were published in the 1950s and 1960s, giving many more details than we will be able to give here. We feel however that since that time much progress has been made in the field and an updated textbook is needed.

In the present volume we try to emphasize the basic physics governing the structure of the stars and the basic ideas on which our understanding of stellar structure is based.

As in the other volumes of this series, we can only discuss the basic principles and leave out the details, sometimes even at the expense of accuracy. We hope to communicate the basic understanding on which further specialized studies can build.

We also want to emphasize the comparison with observations which may support our understanding of stellar evolution or which may show that we still have something to learn.

The book is meant to be a textbook for senior and first-year graduate students in astronomy or physics. We tried to make it understandable for anybody with a basic physics and mathematics education.

We also tried to make this volume understandable for readers who are not familiar with Volumes 1 and 2 of this series. For those readers we give a short introduction which summarizes some basic definitions and facts about stars. Readers who are familiar with the earlier volumes may skip the introduction.

As in the previous volumes, we do not give references for every statement, but rather refer the readers to some of the other textbooks which give the older references. We only give references for the most recent results which are not yet listed in existing textbooks, and for specific data used from other publications.

There are a number of more specialized and more detailed books available. We list a number of these books in the bibliography for those readers who want to learn more about the field than can be presented here.

I am very grateful to Drs K. H. Böhm, W. Brunish, V. Haxton, R. Kippenhahn and J. Naiden for a critical reading of several chapters of this book and for many helpful suggestions.

I am especially indebted to W. Brunish and Ch. Proffitt for supplying a large amount of data and plots which were used for this book.

1

Introduction

1.1 Color magnitude diagrams

1.1.1 Apparent magnitudes of stars

In the first volume of this series we discussed how we can measure the brightnesses of stars, expressed in magnitudes. For a given wavelength band we compare the amount of energy which we receive above the Earth's atmosphere and compare this essentially with the amount that we receive from Vega, also above the Earth's atmosphere. If in this wavelength band the star is brighter than Vega, then its magnitude is smaller than that of Vega. If πf is the amount of energy we receive per cm^2 s from a given star, then the magnitude differences are defined by

$$m_V(1) - m_V(2) = -2.5(\log f(1) - \log f(2)) \qquad (1.1)$$

where (1) refers to star 1 and (2) to star 2. The magnitudes designated by lower case m describe the apparent magnitudes which refer to the energy received here but corrected for the absorption by the Earth's atmosphere (see Volume 1). The subscript V in equation (1.1) indicates measurements in the visual band, i.e. in the wavelength band to which our eyes are sensitive. We can also measure the energy received for star 2 in other wavelength bands (for instance in the blue), compare these with the energy received for star 1 in the same (for instance, blue) wavelength bands, and hence obtain magnitudes for these other wavelength bands. We can thus determine magnitudes in the blue, or in the ultraviolet, etc. We still need to define the magnitudes for star 1, the standard star, for which we use Vega. By definition, all apparent magnitudes, i.e. all lower case m magnitudes for Vega, are essentially zero.

(Strictly speaking, the zero point for the apparent magnitude scale is fixed by the north polar sequence of stars or by the standard star sequence of Johnson and Morgan (1953), but for all practical purposes we can say that the zero point is fixed by Vega. We may make an error in the

1

magnitudes by 0.01 or 0.02, which is about the measuring uncertainty anyway.)

This definition that for Vega $m = 0$ for all wavelengths does not mean that the fluxes received from Vega for different wavelengths are all the same; they are not (see Fig. 1.1).

1.1.2   *The colors*

The stars do not all have the same relative energy distribution as Vega. There are stars which have more flux in the visual spectral region than Vega. Their visual magnitude (V) is therefore smaller than that of Vega. These same stars may have less flux than Vega in the ultraviolet (U) or blue spectral regions (B). In these wavelength bands their magnitudes are larger than those of Vega. The magnitudes of a given star may therefore be different in different wavelength bands. The difference

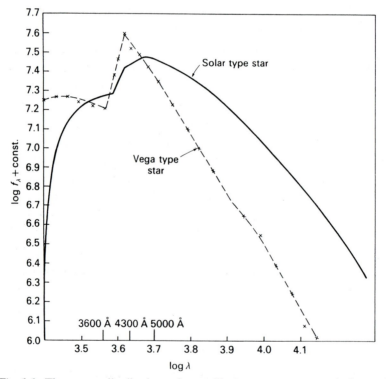

Fig. 1.1. The energy distributions of a star like Vega and of a star similar to the sun. If the solar type star is assumed to have the same apparent brightness in the blue (B) spectral region as the star similar to Vega then it has more light in the visual (V) than the Vega type star. The $m_B - m_V = B - V$ of the solar type star is larger than for Vega, i.e. its $B - V > 0$. It looks more red.

between, for instance, the visual and the blue magnitudes of a given star then tells us something about the energy distribution in the star as compared to the energy distribution for Vega. Stars which have relatively more energy in the visual than in the blue as compared to Vega look more 'red' than Vega. For such stars the difference $m_B - m_V$ is positive (see Fig. 1.1). The difference between the blue and the visual magnitudes is abbreviated by B − V. The difference is called the B − V color. Positive values of B − V mean the star is more 'red' than Vega, negative values of B − V mean the star is more 'blue'. We can define different colors depending on which magnitudes we are comparing. The U − B color compares the magnitude in the ultraviolet and in the blue.

Often the apparent magnitudes m_V, m_B and m_U are abbreviated by V, B and U respectively.

1.1.3 Interstellar reddening

When studying the intrinsic colors and magnitudes of stars we also have to take into account that the interstellar medium between us and the stars may absorb some of the star light. In fact, it is the interstellar dust which absorbs and scatters light in the continuum. It absorbs and scatters more light in the blue and ultraviolet spectral region than in the visual. This changes the colors. The change in color is called the color excess. On average the color excess increases with increasing distance of the star if it is in the galactic plane. Since there is no or very little dust in the galactic halo there is very little additional reddening for stars further out in the galactic halo. In the galactic disk the average change in the B − V color, or the color excess in B − V, is $E(B - V) = 0.30$ per kiloparsec and $\Delta m_V = A_V \sim 1$ magnitude per kiloparsec. Usually the ratio $A_V/E(B - V) \approx 3.2$. The color excess in the B − V colors is larger than in the U − B colors. We find generally $E(U - B)/E(B - V) \approx 0.72$. The apparent magnitudes, corrected for interstellar reddening, are designated by a subscript 0, i.e. m_{BV_0}, m_{B_0}, m_{U_0} etc., and the colors by $(B - V)_0$ and $(U - B)_0$.

1.1.4 The absolute magnitudes of the stars

The amount of energy we receive above the Earth's atmosphere decreases inversely with the square of the distance of the star. If we know the distance of the star we can calculate the apparent magnitude the star would have if it were at a distance of 10 parsec. These magnitudes are called the *absolute magnitudes* and are designated by a capital *M*. We can again determine M_V, M_B, M_U etc., depending on the wavelength band we are considering. We can easily convince ourselves that the colors determined from the absolute magnitudes are the same as those determined from the apparent magnitudes, i.e.

$$m_{B_0} - m_{V_0} = M_B - M_V = B_0 - V_0 = (B - V)_0 \qquad (1.2)$$

Vega is at a distance of 8 parsec (pc). If Vega were at a distance of 10 pc it would be fainter. This means its magnitude would be larger than its actual apparent magnitude. The absolute magnitude of Vega is $M_V = 0.5$.

   Absolute magnitudes tell us something about the intrinsic brightnesses of stars because we compare the brightnesses they would have if they were all at the same distance. For nearby stars the distances d can be determined by means of trigonometric parallaxes, as we discussed in the previous volumes. The relation between apparent and absolute magnitudes is given by

$$m_{V_0} - M_V = 5 \log d(\text{pc}) - 5 \qquad (1.3)$$

where d is the distance measured in parsec (at a distance of 1 parsec, the angular orbital radius of the Earth around the Sun is 1 arcsec). The value of $m_{V_0} - M_V$ is called the distance modulus.

1.1.5   *The color magnitude diagram of nearby stars*

   If we have determined colors and absolute magnitudes of stars, we can plot their positions in a so-called color magnitude diagram, also called a Hertzsprung Russell diagram. (The true Hertzsprung Russell (HR) diagram is actually a spectral type magnitude diagram.) In the color magnitude diagram the absolute magnitudes of the stars are plotted as the ordinate and the colors – usually the $B - V$ colors – as the abscissa. For nearby stars $(E(B - V) \sim 0)$ for which we can measure trigonometric parallaxes, we obtain Fig. 1.2. We find that most stars cluster along a line, the so-called *main sequence*. These stars are also called *dwarf stars*. We also find a few stars which are intrinsically much fainter than main sequence stars. These stars are called *white dwarfs* in order to distinguish them from the 'dwarfs' and also because the first stars discovered in this class were rather bluish or white. Since then, some faint stars with rather reddish colors have been discovered. The name 'white dwarfs' is therefore not always appropriate, but it is nevertheless used for all faint stars of this class in order to distinguish them from the main sequence dwarfs. Since we know several of these faint stars in our neighborhood in spite of the

difficulty of discovering them, because of their faintness, we must conclude that they are rather frequent, but at larger distances we cannot
detect them.

We also see a few stars in our neighborhood which are brighter than
main sequence stars, because they are much larger. These stars are
therefore called *giants*.

1.1.6   Galactic or open clusters

There is of course no reason why we should only compare the
intrinsic brightnesses of stars at a distance of 10 pc; we could just as well
compare them at any other distance – even an unknown distance –

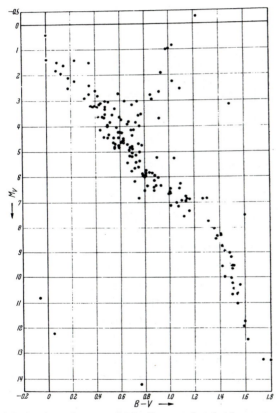

Fig. 1.2. In the color magnitude diagram the absolute magnitudes for nearby stars with
distances known from trigonometric parallaxes are plotted as a function of their colors.
Most of the stars fall along a sequence called the main sequence. A few stars are much
fainter than the main sequence stars also called dwarfs. These very faint stars are called
white dwarfs. We also see some stars brighter than the main sequence stars. These are
called giants. From Arp (1958).

provided only that we are sure that the stars are all at the *same* distance. Such groups of stars are seen in the so-called star clusters. In Fig. 1.3 we show the double star cluster h and χ Persei. The Pleiades star cluster can be seen with the naked eye but is much clearer with a pair of binoculars. These stars are clumped together in the sky and obviously belong together, though some background stars are mixed in with true cluster stars. True cluster stars can be distinguished by their space motion – they must all have approximately the same velocity, because otherwise they would not have stayed together for any length of time. If stars are within a cluster and have the same velocities in direction and speed, we can be quite certain that they are all at the same distance. We can plot color magnitude diagrams for these stars, but we have to plot apparent magnitudes because we do not know their absolute magnitudes. If these stars behave the same way as nearby stars, we would still expect the same kind of diagram because all the magnitudes are fainter by the same constant value, namely,

Fig. 1.3. A photograph of the double star cluster h and χ Persei in the constellation of Perseus. From Burnham (1978a).

$m_{V_0} - M_V = 5 \log d - 5$, where d is the same for all stars. In Figs. 1.4 and 1.5 we show the color magnitude diagrams for the h and χ Persei clusters and for another star cluster, the Praesepe cluster. We can clearly identify the main sequences in these diagrams. In the h and χ Persei clusters we

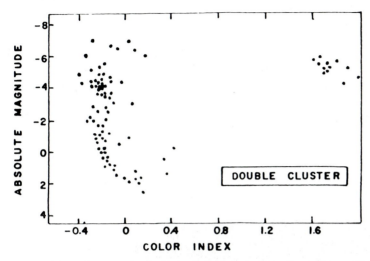

Fig. 1.4. The color magnitude diagram for h and χ Persei star cluster. A distance modulus of $m_{V_0} - M_V = 11.8$ was assumed and $m_V - m_{V_0} = 1.6$. From Burnham (1978b).

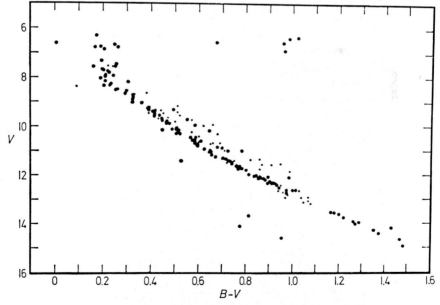

Fig. 1.5. The color apparent magnitude ($m_V = V$) diagram for the stars in the open cluster Praesepe. The stars above the main sequence are probably binaries, and therefore appear brighter. From Arp (1958).

find a few stars which are still brighter than the giants as compared to the main sequence stars. These very bright stars are called the supergiants. The giants and supergiants do not form a very tight sequence as do the main sequence stars. For a given $(B - V)_0$ they show a much larger spread in magnitudes than we find for the main sequence.

1.1.7   Globular clusters

For star clusters similar to those discussed so far which are located in the galactic plane – the so-called open clusters or galactic clusters – we find color magnitude diagrams which look rather similar. There is, however, another group of clusters, the so-called globular clusters. These contain many more stars than open clusters. Many of these globular clusters lie high above the galactic plane. These clusters are characterized by very different color magnitude diagrams. In Fig. 1.6 we show a

Fig. 1.6. A photograph of the globular cluster 47 Tucanae. From Burnham (1978c).

photograph of the globular cluster NGC 104, also called 47 Tucanae or 47 Tuc and in Figs. 1.7 and 1.8 the color magnitude diagrams for 47 Tuc and the globular cluster, called M92, are shown. The globular clusters are all very distant, so all the stars are rather faint. These observations were made only recently and go to very faint magnitudes in comparison for instance with Fig. 1.4. The heavily populated main sequence is clearly recognizable, but only for $M_V > 4$ in M92 and for $m_V = V > 17$ in 47 Tuc. In many globular cluster diagrams we find two branches which go almost horizontally through the diagram. The lower branch is only short, while the upper horizontal branch may extend to quite blue colors and may even turn downward at the blue end (see Fig. 1.8). This upper, extended, horizontal branch is actually called the *horizontal branch*. The lower, stubby, nearly horizontal branch is called the *subgiant branch* because it is brighter than

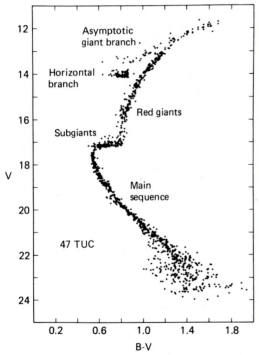

Fig. 1.7. The color apparent magnitude diagram for 47 Tucanae. The new measurements go down to stars as faint as 23rd magnitude, though for the faint stars the scatter becomes large. The main sequence and the giant and subgiant branches are surprisingly sharp, showing that there are very few or no binaries in this globular cluster. The red stub of the horizontal branch is seen at $V \sim 14$ and $B - V \sim 0.8$. The asymptotic giant branch (see Chapter 14) is seen above the horizontal branch. From Hesser *et al.* (1987).

the main sequence but in comparison with the main sequence stars of similar B − V colors not quite as bright as the giant sequence. In addition, we see on the red side an almost vertical sequence. This sequence is called the *red giant branch*.

One of the main aims of this book is to understand why the color magnitude diagrams of these two types of cluster, galactic or open clusters and globular clusters, look so different. In fact, the difference in appearance is mainly due to differences in the ages of the clusters.

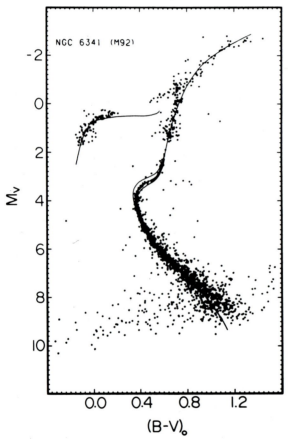

Fig. 1.8. The color absolute magnitude diagram for the globular cluster M92 (cluster 92 in the Messier catalog of nebulous objects). The new observations for M92, like those for 47 Tuc, go to very faint magnitudes. For M92 the main sequence is now clearly recognizable. In addition the subgiant, red giant and horizontal branches are clearly seen. Also seen is the so-called asymptotic branch, for $(B − V)_0 \sim 0.6$ above the horizontal branch. The thin lines shown are the theoretical isochrones, i.e. the location where stars are expected to be seen at a given time. From Hesser *et al.* (1987).

1.2   Stellar luminosities

So far we have talked only about the brightnesses of stars as observed in certain wavelength bands. For the study of stellar structures it is more important to study the total amount of energy radiated by the star per unit of time. Hot stars emit most of their energy at ultraviolet wavelengths, but ultraviolet radiation is totally absorbed in the Earth's atmosphere and can therefore be observed only from satellites. In Fig. 1.9

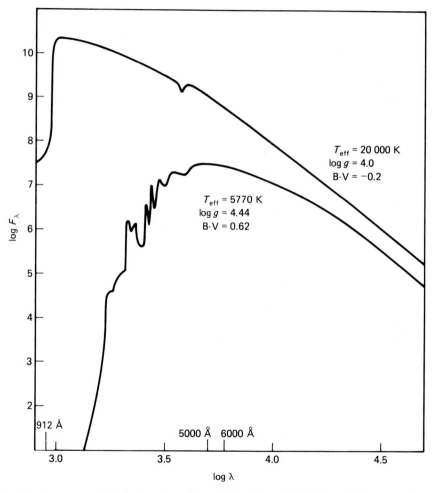

Fig. 1.9. The energy distribution for a star with B − V = −0.2. Most of the energy is emitted in the ultraviolet which cannot be observed from ground but only from satellites. Some energy is emitted at $\lambda < 912$ Å. These short wavelengths are strongly absorbed by interstellar gas. The energy distribution for a star with B − V = 0.62 is also shown. For such stars relatively little energy is emitted in the invisible ultraviolet and infrared spectral regions.

we show the overall energy distribution of a star with $(B - V)_0 = -0.2$, a very blue star. The wavelength range observed through the V filter is indicated. For such hot stars (and there are even hotter ones), a fairly large fraction of energy is emitted in the wavelength region $\lambda < 912$ Å, which is observable only for a very few nearby stars in favorable positions in the sky because radiation at such short wavelengths is generally absorbed by the interstellar medium. The energy distribution in this wavelength region has not yet been well measured for any hot, i.e. very blue, main sequence star. The energy distribution shown in Fig. 1.9 is obtained from theoretical calculations.

In the same diagram we also show the energy distribution for a solar type star. It has its maximum close to the center of the V band. For such stars a relatively small amount of energy is emitted in the ultraviolet and infrared.

In Fig. 1.10 we show the energy distribution for an M4 giant with $(B - V)_0 = 1.5$. For such a star most of the energy is emitted at infrared wavelengths, which can be observed from the ground at certain 'windows', i.e., at certain wavelengths in which the Earth's atmosphere is reasonably transparent, but only with special infrared receivers. Our eyes are not sensitive to these wavelengths.

The total amount of energy emitted per second by a star is called its *luminosity L*. It is measured by the so-called *bolometric magnitudes*, m_{bol} or M_{bol}. Again, we have

$$M_{bol}(1) - M_{bol}(2) = -2.5(\log L(1) - \log L(2)) \qquad (1.4)$$

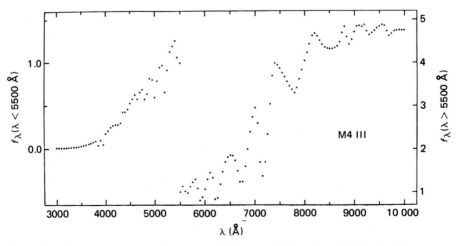

Fig. 1.10. The energy distribution for an M4 giant with $B - V = 1.5$. Most of the energy is emitted at infrared wavelengths. Notice the change of scale at 5500 Å. From Straizys and Sviderskiene (1972).

where (1) refers to star 1 and (2) to star 2. The differences between bolometric and visual magnitudes are called the *bolometric corrections* BC,

$$M_{\text{bol}} = M_V - BC \qquad (1.5)$$

Many astronomers use a different sign in equation (1.5). With the + sign the BC then have to be negative. In any case, bolometric magnitudes are generally smaller than the visual magnitudes because there is more energy in all wavelengths than in a special wavelength band. Equation (1.4) does not completely define the bolometric magnitudes unless we define the bolometric magnitude of star 1. Unfortunately, the zero point is not determined in the same way for the bolometric magnitudes. **The apparent bolometric magnitude for Vega is not zero!**

As we saw above, hot stars emit a large amount of energy in the invisible ultraviolet. They have large bolometric corrections. The cool red stars have a large amount of energy in the infrared and they also have large bolometric corrections. The bolometric corrections have a minimum for stars with $(B - V)_0 \approx 0.35$. For main sequence stars with $(B - V)_0 = 0.35$ the bolometric corrections are *defined* to be zero. With this definition the bolometric magnitude scale is fixed. Using equation 1.5, we find for the sun that $BC(\text{sun}) = 0.07$ and for Vega that $BC(\text{Vega}) \approx 0.3$. Knowing the distance of the sun we find $M_{V_\odot} = 4.82$ and $M_{\text{bol}\odot} = 4.75$.

1.3   Effective temperatures of stars

If we have determined the angular radius of a star, as discussed in Volume 1 of this series, we can determine the amount of energy leaving the stellar surface per cm^2 s into all directions, called the surface flux, πF. We have

$$4\pi R^2 \pi F = L \quad \text{or} \quad \pi F = \frac{L}{(4\pi R^2)} \qquad (1.6)$$

The luminosity of a star can be measured from the flux πf arriving above the Earth's atmosphere per cm^2 s. The total luminosity is given by πf times the surface of the sphere with radius d around the star, where d is the distance of the star,

$$L = \pi f \cdot 4\pi d^2 \qquad (1.7)$$

With equation (1.6) we find

$$L = \pi f \cdot 4\pi d^2 = \pi F \cdot 4\pi R^2 \qquad (1.8)$$

which yields

$$\pi F = \pi f \left(\frac{d}{R}\right)^2 \tag{1.9}$$

where R/d is the angular radius of the star if the angle is measured in radians. The angular radius can be measured for some very bright and several large nearby stars (see Volume 1).

We know that for an ideal light source in complete thermodynamic equilibrium (i.e., for a black body; see Volume 1), the amount of energy emitted per cm^2 s into all directions is

$$\pi F(\text{black body}) = \sigma T^4 \tag{1.10}$$

where T is the temperature of the black body and σ is the Stefan–Boltzmann constant, $\sigma = 5.67 \times 10^{-5}$ erg deg^{-4} cm^{-2} s^{-1}. If we compare the stellar surface flux πF with the radiation of a black body we can define the so-called effective temperature T_{eff} of the star by writing

$$\pi F(\text{star}) = \sigma T_{\text{eff}}^4 \tag{1.11}$$

T_{eff} is then a measure of the surface flux of the star. It is the temperature a black body would need to have in order to radiate the same amount of energy per cm^2 s as the star. In Volume 2 we also saw that T_{eff} is the temperature in the atmosphere of the star at a depth $\bar{\tau} = \frac{2}{3}$, where the optical depth $\bar{\tau}$ is defined as

$$d\bar{\tau} = \bar{\kappa} \, dt \tag{1.12}$$

and where $\bar{\kappa}$ is an average absorption coefficient, averaged over all wavelengths with the Rosseland weighting function (see Chapter 8 of Volume 2). For the sun $\bar{\tau} = \frac{2}{3}$ corresponds to a depth of about 100 km below the Sun's 'surface'.

For the sun we find $\pi f_\odot = 1.38 \times 10^6$ erg cm^{-2} s^{-1} and $m_{\text{bol} \odot} = -26.85$. For $\pi F_\odot$ we derive $\pi F(\text{sun}) = 6.3 \times 10^{10}$ erg cm^{-2} s^{-1} and $T_{\text{eff}}(\text{sun}) = 5800$ K.

1.4   Stellar masses

In Volume 1 we saw that stellar masses can be determined for binary stars. Using the fact that such binary systems are in stable orbits around each other, so that at any moment the gravitational and centrifugal forces are in equilibrium, we can derive that

$$\frac{P^2}{(a_1 + a_2)^3} = \frac{4\pi^2}{G} \frac{1}{(M_1 + M_2)} \qquad (1.13)$$

Here a_1 and a_2 are the semi-major axes of the orbits of stars 1 and 2 around their center of gravity (see Fig. 1.11); M_1 and M_2 are the masses of stars 1 and 2; P is the orbital period; G is the gravitational constant. In order to determine both masses we need to measure the ratio of the orbital velocities v of both stars or the ratio of the semi-major axes. We have

$$\frac{v_1}{v_2} = \frac{M_2}{M_1} \quad \text{or} \quad \frac{a_1}{a_2} = \frac{M_2}{M_1} \qquad (1.14)$$

The orbital velocities can be determined from the Doppler shifts of spectral lines. For a light source moving away from us, the wavelengths λ of the light are shifted by

$$\frac{\Delta \lambda}{\lambda} = \frac{v_r}{c} \qquad (1.15)$$

where c is the velocity of light. The radial velocity v_r is the component of the velocity along the line of sight (see Fig. 1.12). For spectral lines of

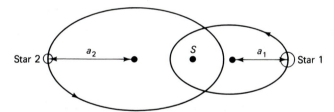

Fig. 1.11. For binary stars both companions orbit the center of gravity. The semi-major axis for star 1 is a_1, that of star 2 is a_2.

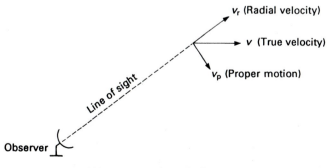

Fig. 1.12. The Doppler shift measures the velocity component along the line of sight, the so-called radial velocity.

Table 1.1. *Data for main sequence stars*

$(B - V)_0$	M_V	M_{bol}	$\log L/L_\odot$	T	$M/M_\odot$
−0.33	−5.7			42 000	(40)
−0.32	−5.0	−8.3	5.2	37 000	23
−0.30	−4.0	−6.8	4.6	30 500	16
−0.25	−2.9	−5.2	4.0	24 000	10
−0.20	−1.5	−4.3	3.6	17 700	7
−0.15	−0.9	−2.1	2.7	14 000	4.5
−0.10	−0.3	−0.9	2.2	11 800	3.6
−0.05	0.3	−0.0	1.9	10 500	3.1
0.00	0.9	0.7	1.6	9500	2.7
0.10	1.5	1.5	1.3	8500	2.3
0.20	2.2	2.2	1.0	7900	1.9
0.30	2.7	2.7	0.8	7350	1.6
0.40	3.3	3.3	0.6	6800	1.4
0.50	3.9	3.9	0.3	6300	1.25
0.60	4.5	4.4	0.1	5900	1.03
0.70	5.2	5.1	−0.1	5540	0.91
0.80	5.8	5.6	−0.3	5330	0.83
0.90	6.3	6.0	−0.5	5090	0.77
1.00	6.7	6.3	−0.6	4840	0.72
1.10	7.2	6.6	−0.7	4590	0.67
1.20	7.5	6.9	−0.8	4350	0.62
1.30	7.9	7.1	−0.9	4100	0.56
1.40	8.8	7.5	−1.1	3850	0.50

known laboratory wavelengths (see Section 1.6), the shifts $\Delta\lambda$ can be measured and thus the radial velocity can be measured. From the radial velocities we can derive the orbital velocities only if we know the inclination of the orbital plane with respect to the line of sight. This inclination, described by the angle i between the normal to the orbital plane and the line of sight, can be determined for eclipsing binaries, where the angle i must be close to 90° (or we would not see eclipses), and for visual binaries when the apparent orbit of one star around the other can be measured.

In Table 1.1 we have collected B − V colors, absolute magnitudes, luminosities, effective temperatures and masses as observed for main sequence stars.

1.5   The mass–luminosity relation

Using the data of Table 1.1, the relation between stellar luminosities and stellar masses is plotted in Fig. 1.13. We know that most stars form a one-dimensional manifold, as indicated by their clustering along the main sequence in the HR diagram. The structure of these stars must be determined by one parameter. The same fact is shown in Fig. 1.13. For a given mass the luminosity is generally nearly the same. The exceptions are the white dwarfs. Nearly all other stars follow the so-called mass–luminosity relation. The surprising fact is that most giants (except the red giants) and supergiants also follow nearly the same mass–luminosity relation as seen for main sequence stars.

In the double logarithmic plot of Fig. 1.13 the points for the stars almost follow a straight line, which means they follow a relation

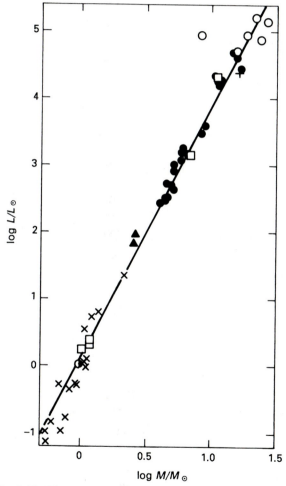

Fig. 1.13. The relation between stellar luminosities and stellar masses for binaries with well determined masses. The different symbols refer to different kinds of binaries. From Popper (1980).

$$L \propto M^\beta \qquad\qquad (1.16)$$

where $\beta \approx 3.8$ on average.

1.6   Spectral classification

If the stellar energy distribution is measured in very small wavelength bands, $\Delta\lambda < 1$ Å, we obtain stellar spectra. They show many narrow wavelength bands with reduced fluxes, the so-called spectral lines. These are due to absorption by atoms and ions in the surface layers of the star, called the stellar atmosphere. Each ion or atom absorbs a characteristic set of lines unique to that particle. From the observed line systems the absorbing particles and elements can be identified, and the element abundances can be determined (see Volume 2). The stellar spectra can be arranged in a sequence according to the lines that are seen. This spectral sequence coincides with the B − V color sequence. The spectra showing the strongest hydrogen lines are called A0 stars. These are stars like Vega and have B − V ~ 0. Stars somewhat more blue, i.e. 0 < B − V < 0.2 are called B stars. The B stars show somewhat weaker hydrogen lines but also show weak helium lines. Even smaller B − V values are found for the so-called O stars. They have weak hydrogen lines, weaker than the B stars, also weak lines of helium, but also ionized helium lines. They are the hottest stars. The O and B stars are called early type stars, because they are at the top of the spectral sequence. Within each spectral class there are subclasses, for instance, B0, B1, . . ., B9. The B0, B1, B2, B3 stars are called early B stars; B7, B8, B9 are late B stars. Stars with $0 \le B − V \le 0.29$ are A stars. Vega is an A0 star, i.e. an early A star. In addition to the strong hydrogen lines, very weak lines of heavy element ions are seen.

Stars with $0.3 < B − V \le 0.58$ are called F stars. Their hydrogen lines are weaker than those for the A stars. Many lines of heavy element ions are seen and these are much stronger than in A stars. For $0.6 < B − V < 1.0$ we find the so-called G stars. The Sun is a G2 star. The hydrogen lines become still weaker for these stars, while the lines of heavy element ions increase in strength. Lines of heavy atoms become visible. For $1.0 < B − V < 1.4$ we find the so-called K stars with weaker lines of heavy ions but increasing line strength for the atomic lines. Their hydrogen lines are very weak. For the red M stars lines of heavy atoms are mainly seen in addition to the molecular bands. Fig. 1.14 shows the sequence of stellar spectra.

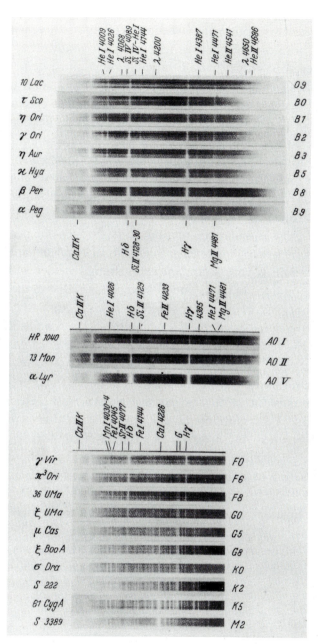

Fig. 1.14. The sequence of stellar spectra. From Unsöld (1955).

1.7   The chemical composition of stars

What are stars made of? In the preceding section we saw that in stellar spectra we see absorption lines whose wavelengths agree with those observed in the laboratory for hydrogen, helium, and many other heavier elements. We therefore know that all these elements must be present in the stars. The question is, what are the relative abundances of these elements? In Volume 2 we discussed how the abundances can be determined. We suspect that the strength of the lines is an indication of the abundance of the corresponding element. It is true that the abundances *can* be determined from line strengths, but the line strengths also depend on temperatures, pressure and turbulence in the stellar atmospheres, which therefore have to be determined together with the abundances. It turns out that hydrogen, for which we see the strongest lines in the A, B and F stars, is the most abundant element in stars. Ninety-one per cent of all heavy particles are hydrogen atoms or ions. About 9 per cent are helium atoms or ions. Helium is very abundant even though we see only weak helium or helium ion lines in the B and O stars. This is due to the special structure of the energy level diagram of helium. It seems then that there is nothing left for the heavier elements which we also see. Indeed the abundance of all the heavier elements together by *number of particles* is only about 0.1 per cent.

The abundance ratios are different if we ask about the *fraction by mass*. The helium nucleus is four times as massive as hydrogen. The heavier nuclei are, of course, even more massive. Carbon has an atomic weight of 12, nitrogen 14, oxygen 16, etc. Iron has an atomic weight of 56. Therefore by mass hydrogen constitutes only about 70 per cent and helium 28 per cent. The mass fraction of hydrogen is usually designated by X and that of helium by Y. We therefore generally have $X = 0.7$ and $Y = 0.28$. The heavier elements together contribute about 2 per cent to the mass. Their abundance by mass fraction is designated by Z, where $Z \simeq 0.02$ for the sun. Even though the heavy elements are very rare in stars it turns out that they are very important for temperature stratification inside the stars, as we shall see.

2

Hydrostatic equilibrium

2.1   The hydrostatic equilibrium equation

What information can we use to determine the interior structure of the stars? All we see is a faint dot of light from which we have to deduce everything. We saw in Volume 2 that the light we receive from main sequence stars comes from a surface layer which has a thickness of the order of 100 to 1000 km, while the radii of main sequence stars are of the order of 10^5 to 10^7 km. Any light emitted in the interior of the stars is absorbed and re-emitted in the star, very often before it gets close enough to the surface to escape without being absorbed again. For the sun it actually takes a photon 10^7 years to get from the interior to the surface, even though for a radius of 700 000 km a photon would need only 2.5 seconds to get out in a straight line. There is only one kind of radiation that can pass straight through the stars – these are the neutrinos whose absorption cross-sections are so small that the chances of being absorbed on the way out are essentially zero. Of course, the same property makes it very difficult to observe them because they hardly interact with any material on Earth either. We shall return to this problem later. Except for neutrinos we have no radiation telling us directly about the stellar interior. We have, however, a few basic observations which can inform us indirectly about stellar structure.

For most stars, we observe that neither their brightness nor their color changes measurably in centuries. This basic observation tells us essentially everything about the stellar interior. If the color does not change it tells us that the surface temperature, or T_{eff}, does not change. If the light output, i.e. the luminosity $L = 4\pi R^2 \sigma T_{\mathrm{eff}}^4$, does not change, we know that the radius R remains constant for constant T_{eff}.

These two facts that T_{eff} and R do not change in time permit us to determine the whole interior structure of the star, as we shall see. Let us first look at the constant radius.

We know that the large interior mass of the star exhibits a large gravitational force on the external layers, trying to pull them inwards. If these layers do not fall inwards there must be an opposing force preventing them from doing so. This force apparently is the same one which prevents the Earth's atmosphere from collapsing, namely the pressure force. In the following we consider the forces working on a column of cross-section 1 cm^2 and of height Δh (see Fig. 2.1). In the direction of decreasing height h, i.e. inwards, we have to consider the gravitational force $F_g = mg$, with g being the gravitational acceleration and m being the mass in the volume under consideration. The mass m equals $\rho \Delta V$, where ρ is the density and ΔV is the volume element which for a cross-section of 1 cm^2 is $\Delta h \times 1 \text{ cm}^2 = \Delta h \text{ cm}^3$. The gravitational force $\vec{F}_g$ is then

$$\vec{F}_g = m\vec{g} = \rho\vec{g}\,\Delta h \tag{2.1}$$

In addition, we have to consider the pressure at the height $h + \Delta h$, i.e. $P(h + \Delta h)$, which is isotropic, but here we consider only the forces on the given volume of gas on which the gas pressure at the height $h + \Delta h$ exerts a downward push. This downward force is $P(h + \Delta h)$. In the opposite direction, pushing upwards from the bottom, is only the pressure at height h, i.e. $P(h)$. Since the volume of gas remains stationary, the opposing forces must be in equilibrium, which means

$$P(h) = \rho g\,\Delta h + P(h + \Delta h) \tag{2.2}$$

or

$$(P(h + \Delta h) - P(h))/\Delta h = -\rho g \tag{2.3}$$

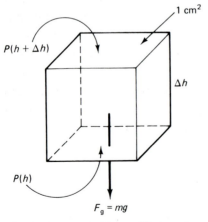

Fig. 2.1. In hydrostatic equilibrium the pressure force dP/dh and the gravitational force F_g working on a volume of gas must balance.

In the limit $\Delta h \to 0$ we have

$$dP/dh = -\rho g \quad \text{or} \quad \text{grad } P = -\rho g \tag{2.4}$$

Equation (2.4) is called the *hydrostatic equilibrium* equation. It expresses the equilibrium between gravitational and pressure forces.

For spherical symmetry we have $dP/dr = -\rho g$.

In the general case other forces may also have to be considered, for instance centrifugal forces for rotating stars or electromagnetic forces in magnetic stars.

We can easily estimate that centrifugal forces have to be taken into account in the surface layers of stars with equatorial velocities of several hundred km s^{-1} as observed for many hot stars, i.e. for stars with $T_{\text{eff}} > 8000$ K. In deeper layers (smaller radii) the centrifugal forces decrease for rigid body rotation. It appears in this case that they do not influence markedly the overall structure of stars.

Magnetic forces may become important for magnetic fields of several thousand gauss as observed in the magnetic peculiar stars, i.e. stars with spectral types Ap or Bp (see Volume 1). For high density stars like white dwarfs or neutron stars only much higher field strengths are of any importance. Magnetic forces are, however, zero for a pure dipole field for which curl $H = 0$. Only for deviations from a dipole field may magnetic forces become non-negligible.

In the following discussion we will consider only gravitational and gas pressure forces unless we specifically mention other forces. This will be adequate for most stars, except for very massive and very luminous stars, for which radiation pressure becomes very important.

From the fact that the vast majority of stars do not shrink or expand we concluded that hydrostatic equilibrium must hold in the stars, as it does in the Earth's atmosphere. Suppose hydrostatic equilibrium did not strictly hold – how fast should we see any effect of the imbalance between pressure and gravitational forces? Maybe it would take so long for the star to change its size that we would not be able to see it. Suppose the equilibrium between pressure and gravitational forces were violated by 10 per cent, so that 10 per cent of the gravitational force is not balanced by the pressure force. This means it would actually be

$$\frac{dP}{dr} = -g\rho + 0.10g\rho$$

Ten per cent of the gravitational force could then pull the material inwards. For a gravitational acceleration of $g_\odot = 2.7 \times 10^4$ cm s^{-2} the net acceleration would then be $g(\text{net}) = 2.7 \times 10^3$ cm s^{-2}. After 1000 seconds, or roughly 15 minutes, the velocity of the material would be

$$\int_0^{100\,\text{s}} g(\text{net})\, dt = 2.7 \times 10^6 \text{ cm s}^{-1} = 27 \text{ km s}^{-1}.$$

The path length s, which the matter would have fallen after 1000 seconds, would be $s = \frac{1}{2}g(\text{net})t^2 = 1.35 \times 10^9$ cm, or 13 500 km. This is a change of nearly 2 per cent of a solar radius within 15 minutes. Such a radius change would become visible very soon. Since we do not see any radius change of most stars after centuries of observation we can be sure that hydrostatic equilibrium must be satisfied to a very high degree of accuracy.

2.2   Consequences of hydrostatic equilibrium

In this section we shall see the conclusions we can draw from the fact that stars must be in hydrostatic equilibrium. First, we will estimate the gas pressure P_g in the center of the sun.

The hydrostatic equation tells us that the pressure in the center of the sun must balance the weight of the overlying material, which is given by the weight of the column of cross-section 1 cm^2 and height R (see Fig. 2.2). With an average density of $\bar{\rho} = 1$ g cm^{-3} this weight is

$$\text{weight} = P_g(\text{center}) = P_{gc} \approx \bar{\rho}\bar{g}R \qquad (2.5)$$

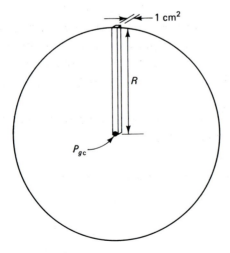

Fig. 2.2. In hydrostatic equilibrium the pressure in the center of the star working on 1 cm^2 must balance the weight of the overlying column of cross-section 1 cm^2 containing a mass $m = \bar{\rho}R$.

or $P_{gc} \approx 4 \times 10^{15}$ dyn cm^{-2}, if we use $\bar{g}_\odot = 2g_\odot = 2 \times 2.7 \times 10^4$ cm s^{-2} which is twice the gravitational acceleration at the solar surface. This indicates a pressure which is at least a billion times greater than that in the Earth's atmosphere.

We now use the equation of state for an ideal gas to obtain an estimate for the central temperature, T_c, of the sun. The equation of state for an ideal gas gives us the relation between P_g, T and ρ, namely

$$P_g = R_g T \rho / \mu \qquad (2.6)$$

where R_g is the gas constant, ρ the density and μ the average atomic weight. $R_g = 8.315 \times 10^7$ [cgs]. Using this equation for the center and making the rough estimate that the central density is about twice the average density, we derive from the central pressure P_{gc} the temperature in the center $T_c = 1.4 \times 10^7$ K, where we have used $\mu = 0.6$ in the center where the material is completely ionized.

Sophisticated model calculations yield $T_c(\text{sun}) = 1.5 \times 10^7$ K. Those same model calculations give $P_{gc} = 10^{17}$ dyn cm^{-2} for the sun, i.e. more than one power of 10 larger than we estimated. The excellent agreement for the central temperatures is therefore somewhat accidental. Our estimate of the central pressure was too low because we used an average gravitational acceleration, which was not the appropriate value. A larger value would have been better. Since we also used a value that was far too low for the central density in estimating the temperature, the errors cancelled. (You might call this the principle of consistent sloppiness which still gives you good results. You have to do it right though!)

The main point is that **we can derive the value for the central temperature by using only the equation of hydrostatic equilibrium**. We have not yet said anything about energy generation or energy transport. We have only used the observed radius of the sun, which is about 700 000 km, and the mass which is 2×10^{33} g.

The second point is the fact that the **central temperatures may be high enough to make nuclear reactions possible**.

2.3 Relation between thermal and gravitational energy: the virial theorem

2.3.1 *Thermal energy*

We saw in the last section that for a given mass M and radius R a certain temperature in the interior is required by hydrostatic equilibrium. If the temperature were lower the pressure would not be high enough to balance the gravitational forces and the star would collapse. If the temperature were higher the pressure would be too great and the star would expand. With the temperature in the star being governed by the hydrostatic equilibrium condition we can compute the thermal energy in the star, if M and R are given. The gravitational energy released during the formation of the star can also be calculated. It is instructive to study the relation between thermal and gravitational energy of a star in hydrostatic equilibrium.

Thermal energy is contained in the gas in the form of kinetic energy $E_{kin} = \frac{3}{2}kT$ per particle, where $k = 1.38 \times 10^{-16}$ erg per degree is the Boltzmann constant.

In addition, energy has to be supplied to ionize the material, which is essentially completely ionized for temperatures of several million degrees. It can, however, easily be estimated that the needed ionization energy is much less than the kinetic energy.

The thermal energy per particle is then given by $E_{kin} = \frac{1}{2}mv^2$. For three degrees of freedom (i.e. motion in three dimensions) this is

$$\tfrac{1}{2}mv^2 = \tfrac{3}{2}kT \tag{2.7}$$

where $m = \mu m_H$ is the average mass of the particles in the gas and m_H is the mass of the hydrogen atom. With $n = \rho/m =$ the number of particles per cm^3 we find for the thermal kinetic energy per cm^3

$$E_{thermal} = n \cdot \tfrac{3}{2}kT \tag{2.8}$$

For each mass shell at distance r from the center and volume $dV = 4\pi r^2\, dr$ it is

$$\Delta E_{thermal} = 4\pi r^2\, dr \cdot n \cdot \tfrac{3}{2}kT \tag{2.9}$$

and the total energy for the whole star comes out to be

$$E_{thermal} = \int_0^R \frac{3}{2}kT \cdot n \cdot 4\pi r^2\, dr \tag{2.10}$$

Making use of the relation for the gas pressure

$$P_g = nkT$$

we can express the thermal energy as

$$E_{thermal} = \int_0^R \frac{3}{2}P_g 4\pi r^2\, dr \tag{2.11}$$

The pressure P_g can be determined from integration of the hydrostatic equation

$$\frac{dP_g}{dr} = -\rho g(r) = -\rho \frac{GM_r}{r^2} \qquad (2.12)$$

where M_r is the mass inside of the sphere with radius r and G is the gravitational constant.

We multiply equation (2.12) by $4\pi r^3$ and integrate by parts from $r = 0$ to $r = R$ and obtain

$$\int_0^R \frac{dP_g}{dr} \cdot 4\pi r^3 \, dr = -\int_0^R \rho \frac{GM_r}{r^2} 4\pi r^3 \, dr = -\int_0^R \rho \frac{GM_r}{r} 4\pi r^2 \, dr \quad (2.13)$$

Integration of the left-hand side by parts gives

$$[P_g \cdot 4\pi r^3]_0^R - \int_0^R 3P_g \cdot 4\pi r^2 \, dr = -\int_0^R 3P_g \cdot 4\pi r^2 \, dr \qquad (2.14)$$

The first term on the left-hand side equals zero because at $r = R$, i.e. at the surface of the star, the pressure $P_g = 0$, and $r^3 = 0$ for $r = 0$. Equation (2.13) then reads

$$-\int_0^R 3P_g 4\pi r^2 \, dr = -\int_0^R \rho \frac{GM_r}{r} 4\pi r^2 \, dr \qquad (2.15)$$

Comparing this with equation (2.11) we see that the left-hand side is twice the thermal energy. We thus find that

$$2E_{\text{thermal}} = \int_0^R \rho \frac{GM_r}{r} 4\pi r^2 \, dr \qquad (2.16)$$

This relation follows quite generally from the hydrostatic equation without our knowing what $P_g(r)$, $\rho(r)$, and $M_r(r)$ are.

2.3.2   Gravitational energy

For the interpretation of the right-hand side of equation (2.15) let us now look at the gravitational energy release during the formation of a star. Considering a starting mass M_r with a radius r at the place of star formation and a mass element $\Delta m = \rho 4\pi r^2\, dr$ falling on M_r from infinity down to the radius r (see Fig. 2.3) the gravitational energy release is

$$\Delta E_G = \int_{\infty}^{r} \text{force } ds$$

which means with the gravitational acceleration $g(s) = GM_r/s^2$

$$\Delta E_G = \int_{\infty}^{r} g(s)\, \Delta m\, ds$$

$$= \int_{\infty}^{r} \frac{GM_r}{s^2}\, \rho 4\pi r^2\, dr\, ds$$

$$= \left[-\frac{GM_r}{s} \right]_{\infty}^{r} \rho 4\pi r^2\, dr$$

$$= -\frac{GM_r}{r}\, \rho 4\pi r^2\, dr \qquad (2.17)$$

Integration over all mass elements building up the star yields

$$E_G = -\int_{0}^{R} \frac{GM_r}{r}\, \rho 4\pi r^2\, dr \qquad (2.18)$$

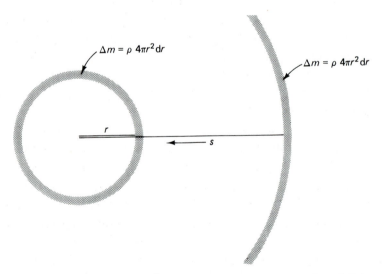

Fig. 2.3. A mass element $\Delta m = \rho 4\pi r^2\, dr$ falls down along the pass s from infinity to the stellar core with mass M_r and radius r.

This expression for the gravitational energy release equals the right-hand side of equation (2.15). In this way we derive the virial theorem, namely

$$E_{\text{thermal}} = -\tfrac{1}{2}E_{\text{G}} \qquad (2.19)$$

Since we did not make any approximation or assumptions (except for $E_{\text{thermal}} = E_{\text{kin}}$), **this equation holds exactly for any $P_g(r), T(r), M_r(r)$ as long as the star is in hydrostatic equilibrium, and as long as we have an ideal gas.**

2.4   Consequences of the virial theorem

Equation (2.19) has very important consequences. For instance, let us look at the early stages of stellar evolution. When the star starts to contract it releases gravitational energy which is transformed into thermal energy and the star heats up. It reaches hydrostatic equilibrium when one half of the released gravitational energy is stored as thermal energy. If more than one half of the gravitational energy were stored as thermal energy the star would be too hot in the interior, the pressure would then be too high, the pressure forces would be larger than the gravitational forces and the star would have to expand again. Due to the expansion and the consumption of gravitational energy the star would cool off until equilibrium is again reached. We see that before the star can contract further it has to lose some of its thermal energy by radiation at the surface. If no other energy source than gravitational energy is available, the temperature would decrease, the star can then continue to collapse, heating up in the process. After each infinitesimal step of collapse the star has to wait until it has radiated away half of the released gravitational energy before it can continue to contract. With increasing gravitational energy release during contraction, i.e. with decreasing radius E_{thermal} has to increase in order to balance the increasing gravitational forces, the star continues to heat up while the radiative energy loss at the surface tries to cool it off.

How long does it take the star to contract to its main sequence size? Long enough to allow the excess heat to leave the star. If the heat transport from the inside out is mainly by radiation then the star has to wait until the radiation has found its way out from the interior. How long does this take? We have seen, for instance, that the sun loses each second an amount of energy given by its luminosity, $L = 4 \times 10^{33}$ erg s^{-1}. The total gravitational energy release of the sun is given by (2.18). For a star with constant density we derive with $M_r = \tfrac{4}{3}\pi r^3 \rho$ that

$$E_G = -\int_0^R \rho^2 \tfrac{16}{3}\pi^2 Gr^4 \, \mathrm{d}r = -\rho^2 G \tfrac{16}{15}\pi^2 R^5 = -\frac{3}{5}\frac{GM^2}{R} \qquad (2.20)$$

Inserting the values for the solar mass and solar radius $M_\odot = 2 \times 10^{33}$ g and $R_\odot = 7 \times 10^{10}$ cm we find

$$E_{G_\odot} = \frac{2.4 \times 10^{66}}{7 \times 10^{10}} G = 2.4 \times 10^{48} \text{ erg}$$

One half of this had to be lost by radiation while the Sun contracted. Assuming that it was shining at the same rate as it is now, we find for the contraction time

$$t = \frac{1.2 \times 10^{48}}{4 \times 10^{33}} \, \mathrm{s}^{-1} = 3 \times 10^{14} \, \mathrm{s}$$

or 10^7 years. It must have taken the sun approximately 10 million years to contract. **This contraction time for stars is generally called the Kelvin–Helmholtz time.**

We can look at this problem from a very different angle.

The energy is liberated mainly in the stellar interior with the large mass concentration. The energy then has to get out by means of photons traveling to the surface. Let us look at the example of the sun, which has a radius of $700\,000$ km $= 7 \times 10^{10}$ cm. If the photons could escape freely they would reach the surface in 2.5 seconds. Actually, they are absorbed after a very short distance, namely one photon mean free path λ_p. The absorbing electron stays in an excited energy level for about 10^{-8} seconds and then re-emits the photon in an arbitrary direction. Soon the photon is re-absorbed, then re-emitted again, etc., proceeding in a random walk process until it finally gets to the surface (see Fig. 2.4). How long does this take on average? We first must know the mean free path. As a rough

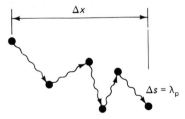

Fig. 2.4. The photon proceeds in a random walk process to the stellar surface. After it has traveled on average a distance Δs of one mean free path λ_p it is absorbed and reemitted in an arbitrary direction. After z absorption and reemission processes it has on average traveled in a given direction x a distance $\Delta x = \sqrt{\bar z} \cdot \lambda_p$.

estimate we find the mean free path to be about 0.5 cm: The absorption cross-section per particle is roughly 2×10^{-24} cm^2, which means in a column with 1 cm^2 cross-section, 5×10^{23} particles cover the whole column cross-section of 1 cm^2. The average density $\bar{\rho}$ for the Sun is $\bar{\rho} \approx 1$ g cm^{-3}. This means on average there are 10^{24} particles per cm^3 ($\mu \sim 0.6$ for ionized material). The projected cross-sections of a column of length 0.5 cm of these particles is 1 cm^2. This means that after the photon has proceeded 0.5 cm it has certainly experienced one absorption process. It can proceed just 0.5 cm before being absorbed, or the mean free path $\lambda = 0.5$ cm. The question then is: how often is the photon absorbed and re-emitted before it reaches the surface?

In Fig. 2.4 we show a short stretch of a random walk process. Statistics tell us that on average the distance Δx travelled in one direction is given by

$$\Delta x = \sqrt{z}\,\lambda_p \qquad (2.21)$$

where z is the number of absorption and re-emission processes. The photon reaches the surface for $\Delta x = R = 7 \times 10^{10}$ cm. In order for the photon to escape we require

$$R = \sqrt{z}\,\lambda_p \quad \text{or} \quad \sqrt{z} = R/\lambda_p = 1.4 \times 10^{11} \quad \text{or} \quad z \approx 2 \times 10^{22}$$

Each re-emission process takes about 10^{-8} s, which means with 2×10^{22} absorption and re-emission processes the travel time for the photon is about 2×10^{14} s or roughly 10^7 years (1 year is about 3×10^7 seconds).

The photons need 10^7 years to reach the surface. This time determines the contraction times of the stars. For more massive stars we find that M/R is approximately constant. (Actually, it increases somewhat for increasing mass.) With this we find $\bar{\rho} \propto R^{-2}$ and $\lambda_p \propto \rho^{-1} \propto R^2$ if the average absorption cross-section is approximately independent of temperature. We would then find $\sqrt{z} \propto R/\lambda_p \propto R^{-1}$ or the contraction time $t \propto z \propto R^{-2}$. If $R \sim 10\,R_\odot$ the contraction time would be 100 times shorter than that for the sun. For more massive stars the contraction times are shorter because the average density is lower and the photons can escape faster even though the radii are larger.

3

Thermal equilibrium

3.1 Definition and consequences of thermal equilibrium

As we discussed in Chapter 2, we cannot directly see the stellar interior. We see only photons which are emitted very close to the surface of the star and which therefore can tell us only about the surface layers. But the mere fact that we see the star tells us that the star is losing energy by means of radiation. On the other hand, we also see that apparent magnitude, color, T_{eff}, etc., of stars generally do not change in time. This tells us that, in spite of losing energy at the surface, the stars do not cool off. The stars must be in so-called *thermal equilibrium*. If you have a cup of coffee which loses energy by radiation, it cools unless you keep heating it. If the star's temperature does not change in time, the surface layers must be heated from below, which means that the same amount of energy must be supplied to the surface layer each second as is taken out each second by radiation.

If this were not the case, how soon would we expect to see any changes? Could we expect to observe it? In other words, how fast would the stellar atmosphere cool?

From the sun we receive photons emitted from a layer of about 100 km thickness (see Volume 2). The gas pressure P_g in this layer is about 0.1 of the pressure in the Earth's atmosphere, namely, $P_g = nkT = 10^5$ dyn cm^{-2}, where $k = 1.38 \times 10^{-16}$ erg deg^{-1} is the Boltzmann constant, T the temperature and n the number of particles per cm^3. From this we calculate that n is about 10^{17} particles per cm^3. The number of particles in a column of 1 cm^2 cross-section and 100 km height is then 10^7 cm $\times 10^{17}$ cm$^{-3} = 10^{24}$ particles per cm^2. With an average temperature of 6000 K in the layer, the energy per particle is roughly $\frac{3}{2}kT = 1.5 \times (1.38 \times 10^{-16}) \times (6 \times 10^3) \approx 10^{-12}$ erg. In the column of 1 cm^2 cross-section from which the photons escape we have an energy content of $E = 10^{-12} \times 10^{24}$ erg $= 10^{12}$ erg (see Fig. 3.1). The energy loss per cm^2 s is given by the surface flux $\pi F = \pi F_{\odot} \approx 6 \times 10^{10}$ erg cm^{-2} s^{-1}. This means that if the sun keeps shining at this rate the total energy content of the atmosphere would be emitted in $t \approx E/\pi F \approx 15$ seconds. We certainly should see changes in the temperature and radiation.

We can then conclude that the heating of the surface layers from below must keep pace with the cooling. It must supply as much energy as is lost at the surface. As the surface flux remains constant, so must the rate of heating. The layer below the surface which supplies the heating cannot change either; it cannot cool or it could not keep up the heat supply. This means it also must be heated by a constant rate, and the same must hold for the next deeper layer, etc. Of course, the deeper we go the longer is the time after which we would see changes if they occurred, because it takes the photons longer before they get to the surface. We would see changes in the heat transport in the interior of the sun only after 10^7 years.

For most stars we do not see any changes in the surface flux. We must therefore conclude that the energy loss at the surface is replaced from below. In a plane parallel layer this means dF/d$z = 0$, where z is the depth below the surface (Fig. 3.2). In spherical geometry this means (see Fig. 3.3)

$$\frac{\mathrm{d}}{\mathrm{d}r}(r^2 \pi F) = 0 \quad \text{or generally} \quad \mathrm{div}\, F = 0 \quad \text{or} \quad \frac{\mathrm{d}L}{\mathrm{d}r} = 0 \qquad (3.1)$$

if there is no energy generation; πF always indicates the total energy flux per cm^2 s. Here r is the distance from the center and L is the luminosity of

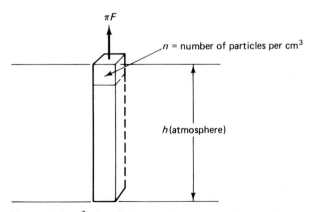

Fig. 3.1. For each 1 cm^2 of surface area the amount of energy loss per second is πF. This energy is taken out of a column of the atmosphere with 1 cm^2 cross-section and height h, in which the number of particles is nh, and in which the total amount of energy is $E = \frac{3}{2}kTnh$.

the star, i.e. $L = 4\pi R^2 \pi F$, where R is the radius of the star; πF is then increasing with depth proportional to r^{-2}. At the surface the energy transport must generally be by radiation, i.e. $\pi F = \pi F_r$, where F_r is the radiative energy flux – otherwise, matter would have to escape from the star (only a minute fraction of stellar energy loss is due to stellar winds). In the deeper layers of the stars the energy transport can be by other means also, i.e. $F = F_r + \cdots$. As we saw in Volume 2 convective energy flux, F_c, may have to be considered. The energy flux due to heat conduction, F_{cd}, can become important only if the mean free path of ions and atoms becomes very large. In special cases, such as in the white dwarfs, we have to consider this possibility.

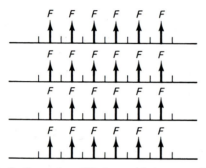

Fig. 3.2. In thermal equilibrium and plane parallel geometry the same amount of energy flux must go through each horizontal layer in every depth.

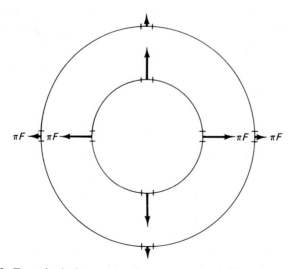

Fig. 3.3. For spherical geometry the same amount of energy must go through each spherical shell. The flux F must increase proportional to r^{-2}.

Equation (3.1) is a necessary (and sufficient) condition if the temperature is to remain constant through the star provided there is no energy source in a given layer. We call this the condition of **thermal equilibrium**. In the special case when the energy transport is only by radiation we call it **radiative equilibrium**. In this latter case equation (3.1) becomes

$$\frac{d}{dr}(r^2 F_r) = 0 \tag{3.2}$$

This is the equation for radiative equilibrium, provided that no energy generation takes place. If energy is generated in a layer then this additional energy has to be transported out so the layer is not heated by the additional energy supply. The energy flux must therefore increase by the amount of energy which is generated, i.e.

$$\text{div } \pi F = \varepsilon \rho \tag{3.3}$$

or for spherical symmetry

$$\frac{dL_r}{dr} = 4\pi r^2 \varepsilon \rho \tag{3.4}$$

where ε is the energy generated per gram of material per second and ρ is the density; $\varepsilon \rho$ is the energy generated per cm^3 s. Equation (3.3) determines the depth dependence of the energy flux. The thermal energy flux determines the temperature stratification, as we saw in Volume 2. The heat flow always goes in the direction of decreasing temperature. The steeper the temperature gradient, the larger the heat flux. A prescribed heat flow πF can only be achieved with *one* temperature gradient. In the following we shall briefly discuss the relation between the energy flux and the temperature gradient, especially the relation between the radiative energy flux πF_r and the temperature gradient.

3.2   Radiative energy transport and temperature gradient

In Volume 2 we derived the relation between the radiative energy flux πF_r and the temperature gradient dT/dr. As for any heat energy transport we expect a larger heat flux for a steeper temperature gradient. We also expect a smaller energy flux if the transport is made difficult. For heat transport by radiation, i.e. by photons, heat transport becomes difficult if the photons are frequently absorbed and re-emitted, in other words if the mean free path for photons becomes very small. This, of course, happens when the absorption coefficient becomes large.

An accurate derivation of the relation between radiative flux and temperature gradient for the plane parallel case is given in Volume 2. In Appendix A we summarize the derivation for the spherically symmetric case in stellar interiors.

Qualitatively we can see the relation between radiative flux and temperature gradient in deep layers when we consider the radiative energy flowing through a cross-section of $1 \, \mathrm{cm}^2$ in a horizontal layer (see Fig. 3.4). In a given layer at depth z_0 we find photons propagating outwards which have been emitted at different depths z but on average originated at a layer where $z = z_0 + \frac{2}{3}\lambda_\mathrm{p}$ below depth z_0 (if they were all propagated radially outwards they would come from $z = z_0 + \lambda_\mathrm{p}$). The amount of energy emitted at this layer is given by the Planck function $\pi F_\mathrm{u} = \pi B = \sigma T^4$ ($\sigma = $ Stefan–Boltzmann constant). (For justification see Appendix A.) The amount of energy propagating outwards is therefore given by $\pi B(z_0 + \frac{2}{3}\lambda_\mathrm{p})$. Propagating downwards are the photons which on average have been emitted at $z = z_0 - \frac{2}{3}\lambda_\mathrm{p}$. The energy propagating downward through $1 \, \mathrm{cm}^2$ at z_0 is given by $\pi F_\mathrm{d} = \pi B(z_0 - \frac{2}{3}\lambda_\mathrm{p})$. The net energy transport outward through $1 \, \mathrm{cm}^2$ at z_0, the $\pi F_\mathrm{r}(z_0)$ is given by the difference $\pi F_\mathrm{u} - \pi F_\mathrm{d}$. We now consider a linear depth dependence of the Planck function B,

$$B\left(z_0 + \frac{2}{3}\lambda_\mathrm{p}\right) = B(z_0) + \frac{2}{3}\frac{\mathrm{d}B}{\mathrm{d}z}\lambda_\mathrm{p} \quad \text{and} \quad B\left(z_0 - \frac{2}{3}\lambda_\mathrm{p}\right) = B(z_0) - \frac{2}{3}\frac{\mathrm{d}B}{\mathrm{d}z}\lambda_\mathrm{p}$$

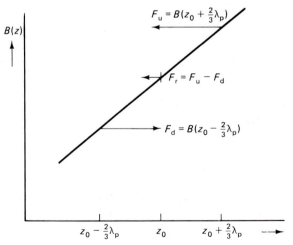

Fig. 3.4. At depth z_0 the photons going outwards originate on average at depth $z_0 + \frac{2}{3}\lambda_\mathrm{p}$. Those going downwards come on average from a depth $z_0 - \frac{2}{3}\lambda_\mathrm{p}$. The net radiative flux F_r is given by $F_\mathrm{r} = F_\mathrm{u} - F_\mathrm{d}$.

We then find

$$F_r(z_0) = B\left(z_0 + \frac{2}{3}\lambda_p\right) - B\left(z_0 - \frac{2}{3}\lambda_p\right) = \frac{4}{3}\frac{dB}{dz}\lambda_p \tag{3.5}$$

How large is the mean free path of photons? Each atom has an absorption cross-section $\kappa_{at} = \kappa$ per atom (dimension cm^2). If there are n atoms per cm^3 the total cross-section of these atoms is $n\kappa_{at} = \kappa$ per cm^3, abbreviated by κ_{cm} (dimension cm^{-1}). If we consider a column of length l for which the whole cross-section is covered by the projection of the photon-absorbing atoms then the photon will certainly be absorbed when passing through this column of length l. The length l is given by $l = 1/\kappa_{cm}$ and is the mean free path λ_p. Inserting $\lambda_p = 1/\kappa_{cm}$ into equation (3.5) we find

$$F_r = F_u - F_d = \frac{4}{3}\frac{1}{\kappa_{cm}}\frac{dB}{dz} = \frac{4}{3}\frac{dB}{d\tau} \tag{3.6}$$

where we have introduced the new variable τ, called the optical depth. It is $d\tau = \kappa_{cm}\,dz$.

With $B = \sigma T^4/\pi$ we find

$$F_r = \frac{\sigma}{\pi}\frac{16}{3}T^3\frac{1}{\kappa_{cm}}\frac{dT}{dz} \tag{3.7}$$

which gives the **relation between the temperature gradient and the radiative flux**. (F_r is considered to be positive when directed outwards.)

Equation (3.7) can be solved to give the temperature gradient for any given value of F_r. We find with $d/dz = -(d/dr)$

$$\frac{dT}{dr} = -\frac{\pi F_r}{\sigma}\frac{3}{16}\frac{\kappa_{cm}}{T^3} \tag{3.8}$$

The total radiative energy transport through a spherical shell with radius r is given by $L_r = 4\pi r^2 \cdot \pi F_r$. We therefore can also write

$$\frac{dT}{dr} = -\frac{L_r}{4\pi r^2}\frac{3}{16}\frac{\kappa_{cm}}{\sigma T^3} \tag{3.9}$$

This determines the temperature stratification for a given radiative energy transport. For radiative equilibrium we find of course $L_r = L$, the luminosity at radius r, which means

$$\frac{dT}{dr} = -\frac{L}{4\pi r^2}\frac{3}{16}\frac{\kappa_{cm}}{\sigma T^3} \tag{3.10}$$

3.3   A first approximation for the mass–luminosity relation

It is quite instructive to see that with the help of the hydrostatic and radiative heat transport equations we can already understand qualitatively the observed mass–luminosity relation (see Chapter 1)

$$L \propto M^m \quad \text{with} \quad 3 < m < 4 \tag{3.11}$$

even without any knowledge about the origin of the energy needed to maintain L. The mere fact that most stars are stable tells us that such a relation must exist. The argument goes as follows: generally we can say that on average the density ρ must be proportional to the mass divided by the volume of the star, i.e.

$$\rho \propto \frac{M}{R^3} \tag{3.12}$$

Hydrostatic equilibrium requires

$$\frac{dP_g}{dr} = -g\rho = -\frac{GM_r}{r^2}\rho \tag{3.13}$$

which can be roughly approximated by

$$\frac{P_g}{R} \approx \frac{GM}{R^2}\frac{M}{R^3} \quad \text{or} \quad P_g \propto \frac{M^2}{R^4} \tag{3.14}$$

Making use of the equation of state $P_g = R_g T\rho/\mu$ we find for the temperatures

$$T \propto \frac{P_g}{\rho} \propto \frac{M^2}{R^4}\frac{R^3}{M} \propto \frac{M}{R} \tag{3.15}$$

Inserting this into equation (3.10) and solving for the luminosity, we derive for radiative equilibrium, i.e. $L = L_r$, that

$$L \propto -R^2 T^3 \frac{dT}{dr}\frac{1}{\kappa_{cm}} = -\frac{T^3}{\kappa_g\rho}\frac{dT}{dr}R^2 \tag{3.16}$$

Here we have replaced κ per cm, κ_{cm}, by κ per gram of material, κ_g, i.e. the absorption coefficient for a column of length $s = 1/\rho$ which contains 1 gram of material. Since a column with 1 cm^2 cross-section and length 1 cm contains ρ grams of material, κ_{cm} must then be $\kappa_g\rho$. The absorption coefficient κ_g is less dependent on pressure and density than κ_{cm} because one gram of material always contains nearly the same number of particles, n, which can absorb light. Changes of n can only be due to changing degrees of ionization. Therefore, it is often better to use κ_g in these discussions. Making here the rough approximation that κ_g is always the same, we find, using equations (3.12) and (3.15),

$$L \propto \frac{M^3}{R^3} \frac{R^3}{M} \frac{M}{R^2} R^2 \tag{3.17}$$

where we have replaced dT/dr by $-T_c/R$. Equation (3.17) then yields

$$L \propto M^3 \tag{3.18}$$

This is the mass–luminosity relation corresponding to the observed relation (3.11). Here we derived $m = 3$ if κ_g is independent of P_g and T and $P \sim P_g$. Our calculations tell us that the **temperature in the interior is** determined by hydrostatic equilibrium as we saw earlier, and that this also **determines the luminosity**. With the temperature at the surface being close to zero the average temperature gradient $dT/dr \sim -T_c/R$ is given. Once κ is known, **the radiative heat transport determines** how much energy is transported to the surface, and this is of course **the luminosity**. As is seen from equation (3.16) the larger the central temperature and the smaller the κ, i.e. the easier the heat transport, the larger will be the luminosity.

3.4 Energy transport by heat conduction

If in a given gas we have a temperature gradient, then particles from the high temperature region with their high thermal energies and high thermal velocities fly into the lower temperature region carrying kinetic energy into it. At the same time lower energy particles with lower velocities fly from the low temperature region into the higher temperature gas, but these particles carry less energy into the higher temperature layers than those moving in the opposite direction, the difference depending on the temperature gradient. There is therefore a net energy transport from the higher temperature gas to the lower temperature gas. The amount of energy transport is directly proportional to the temperature gradient. Therefore, we can say that the thermal motions of the particles lead to a conductive energy flux πF_{cd} which is given by

$$\pi F_{cd} = -\eta \frac{dT}{dr} \tag{3.19}$$

where η is the heat conduction coefficient. The negative sign indicates that the heat flux goes in the direction of decreasing temperature.

The total heat flux is then

$$\pi F = \pi F_{\mathrm{r}} + \pi F_{\mathrm{cd}} = -\frac{1}{\kappa_{\mathrm{cm}}}\frac{16}{3}\sigma T^3 \frac{\mathrm{d}T}{\mathrm{d}r} - \eta \frac{\mathrm{d}T}{\mathrm{d}r} \qquad (3.20)$$

or

$$\pi F = -\frac{\mathrm{d}T}{\mathrm{d}r}\frac{16}{3}\sigma T^3 \left(\frac{1}{\kappa_{\mathrm{cm}}} + \eta \frac{3}{16\sigma}\frac{1}{T^3}\right) = -\frac{\mathrm{d}T}{\mathrm{d}r}\frac{16}{3}\sigma T^3 \left(\frac{1}{\kappa_{\mathrm{cm}}} + \frac{1}{\kappa_{\mathrm{hc}}}\right) \qquad (3.21)$$

where $\kappa_{\mathrm{hc}} = \frac{16}{3}\sigma T^3/\eta$.

This means we can express the conductive flux F_{cd} in analogy to the radiative energy flux by means of

$$\pi F_{\mathrm{cd}} = -\frac{1}{\kappa_{\mathrm{hc}}}\frac{16}{3}\sigma T^3 \frac{\mathrm{d}T}{\mathrm{d}r} \qquad (3.22)$$

and the total flux πF as

$$\pi F = -\frac{1}{\bar{\kappa}}4\sigma T^3 \frac{\mathrm{d}T}{\mathrm{d}r} \qquad (3.23)$$

with

$$\frac{1}{\bar{\kappa}} = \frac{1}{\kappa_{\mathrm{cm}}} + \frac{1}{\kappa_{\mathrm{hc}}} \qquad (3.24)$$

We can thus take heat conduction into account by including it in the definition of $\bar{\kappa}$ which has to be calculated as the harmonic mean of the radiative absorption coefficient κ_{cm} and the 'conductive absorption coefficient' κ_{hc}. If the heat conduction is large, i.e. if η is large, then κ_{hc} is very much smaller than κ_{cm} and $\bar{\kappa}$ is determined by heat conduction. If η is small then κ_{hc} is large and κ_{hc} can be neglected with respect to $\kappa_{\mathrm{cm}}^{-1}$. $\bar{\kappa}$ is then equal to κ_{cm}. η is large if the mean free path of the particles is large such that the higher energy particles can fly far into the lower temperature range. The mean free path λ_{f} for particle collisions in a gas is given by

$$\lambda_{\mathrm{f}} = \frac{1}{Qn} \qquad (3.25)$$

where n is the number of particles per cm^3 and Q is the collisional cross-section. (λ_{f} is again the length of a column over which the sum of all cross-sections, i.e. nQ, for a column of $1\ \mathrm{cm}^2$ cross-section, covers the whole column cross-section.) In the outer layers of the sun $n \approx 10^{17}\ \mathrm{cm}^{-3}$. For collisions between neutral particles we can very roughly estimate that the collisional cross-section is of the order of the atomic cross-section, i.e. about 10^{-16} cm^2. We then find $\lambda_f = (10^{-16} \times 10^{17})^{-1}$ cm $= 0.1$ cm. For charged particles like the electrons the cross-section becomes larger and the mean free path becomes still smaller.

In Chapter 2 we saw that the mean free path for photons, namely κ_{cm}^{-1}, in the solar atmosphere is of the order of 10^7 cm or larger and may be a factor of 10 to 100 times smaller or larger in other atmospheres. Heat transport by photons is therefore generally much more efficient than heat transport by heat conduction. Exceptions are gases with low particle densities and high temperatures, as in the solar corona, where the mean free path for collisions becomes very large. At the same time, the photon density becomes very low in the corona as compared to σT^4. Radiative energy transport therefore is very small in the corona. In these high temperature regions heat conduction becomes important. Another exception is a very high density gas, when the electrons are packed so densely that the Pauli principle becomes important (see Section 14.3); the mean free path of the electrons can then become so large that heat conduction is the main mechanism of energy transport.

4

The opacities

4.1 Bound-free and free-free absorption coefficients

If we want to use equation (3.7) to calculate the temperature gradient we have to determine the absorption coefficient κ.

We discussed in Volume 2 different absorption processes. In Fig. 4.1 we show a schematic energy level diagram for an atom or ion. Absorption in spectral *lines* takes place if an electron, bound in an atom with a number of discrete energy levels, makes a transition from one bound energy level to another bound energy level. The energy difference between these two levels equals the energy $h\nu$ of the absorbed photon. The transitions

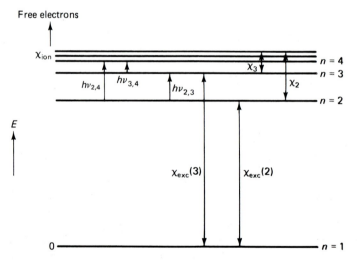

Fig. 4.1. A schematic energy level diagram corresponding to the hydrogen atom. The excitation energy above the ground level, $\chi_{\text{exc}}(n)$, is given by $\chi_{\text{ion}} - \chi_n$. Here χ_{ion} is the energy needed to remove an electron in the ground level from the atom. For hydrogen atoms $\chi_n = \chi_{\text{ion}}/n^2$ and $\chi_{\text{exc}}(n) = \chi_{\text{ion}} - \chi_n = \chi_{\text{ion}}(1 - n^2)$. For the electron to make a transition from the energy level with main quantum number $n = 2$ to the level with $n = 3$ the absorbed photon must have an energy $h\nu_{23} = \chi_2 - \chi_3$. This leads to an absorption of a given frequency, i.e. to line absorption.

between these discrete energy states are called bound-bound transitions because both energy states refer to electrons that are bound to an atom or ion. Such transition leads to absorption in essentially one frequency, the frequency of a particular spectral line. In this section we will discuss the absorption coefficients in the continuum of frequencies extending over all frequencies. Such a *continuous* absorption can only occur when either the energy state from which the absorption starts or the final energy state is not quantized, such that the electron can have a continuum of energy states as initial state or that it may go into a continuum of energy states after the absorption (see Fig. 4.2). Such a continuum of energy states exists only for free electrons. We therefore find continuous absorption only if the final energy level or the starting energy level or both correspond to a free electron. If only the final value is unbound the transitions are called bound-free transitions; if both levels are unbound the transitions are called free-free transitions.

4.1.1   The bound-free absorption coefficients

The bound-free transitions correspond to an ionization process: a bound electron transfers from the bound state into a free state. The inverse

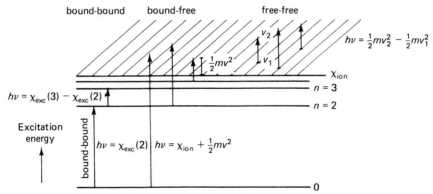

Fig. 4.2. For an absorption process in a continuum of frequencies the electron either has to be in a continuous set of energy levels or it has to transfer into a continuous set of energy levels. Kinetic energies constitute a continuum of energy levels. For an absorption process in a continuum of frequencies the electron must therefore be a free flying electron or it has to become a free electron in the absorption process. In the latter case the electron must be removed from the atom. The atom is ionized. The frequency of the absorbed photon is given by $h\nu = \chi_n + \frac{1}{2}mv^2$, if the electron is removed from the level with main quantum number n. $\frac{1}{2}mv^2$ is the kinetic energy of the electron after the ionization (absorption) process. Such transitions from bound levels are called bound-free transitions. A free flying electron with kinetic energy $\frac{1}{2}mv_1^2$ can also absorb a photon of energy $h\nu$ and obtain a higher kinetic energy $\frac{1}{2}mv_2^2$. The energy of the absorbed photon is given by $h\nu = \frac{1}{2}mv_2^2 - \frac{1}{2}mv_1^2$. It can have a continuous set of values.

process coupled with the emission of a photon is a recombination process. From Fig. 4.2 it is obvious that for such a bound-free transition from a given level with quantum number n a minimum energy of $h\nu = \chi_{ion} - \chi_{exc}(n) = \chi_n$ is required to remove the electron from the atom. Here χ_{ion} is the energy necessary to move an electron in the ground level into the continuum. This is called the ionization energy. $\chi_{exc}(n)$ is the energy difference between the energy level with quantum number n and the ground level. For the hydrogen atom, for instance, with an ionization energy $\chi_{ion} = 13.6$ eV the bound-free absorption from the level with $n = 2$ can only occur if the energy of the photon is larger than $\chi_n = \chi_{ion}/n^2 = 3.4$ eV. This means the wavelength of the absorbed photon has to be shorter than that corresponding to $h\nu = 3.4$ eV, i.e. shorter than 3647 Å.

Such bound-free transitions are possible for all atoms or ions which still have an electron bound to them. (We cannot have bound-free transitions from a proton.) In order to calculate the total bound-free absorption coefficient we first have to calculate the absorption coefficient for each bound level in each kind of atom or ion as a function of frequency, then total all the contributions at each frequency. Generally this absorption coefficient is wavelength dependent. We therefore have to decide which absorption coefficient we have to use in equation (3.7) if we want to calculate the temperature stratification. Since the temperature gradient is determined by the amount of radiative flux which can be transported through a given horizontal layer, the most important wavelength band is the one in which most of the flux is transported. This wavelength band is determined by the wavelengths in which most of the energy is emitted. With $F_r \propto \kappa^{-1}(dB/dz)$ (equation (3.7)) this means mainly by the wavelength for the maximum of the Planck function and by the minimum of κ since $F_r \propto \kappa^{-1}$. For higher temperatures the maximum of the Planck function shifts to shorter wavelengths. Requiring that

$$\int_0^\infty F_\lambda \, d\lambda = \int_0^\infty \frac{4}{3} \frac{1}{\kappa_\lambda} \frac{dB_\lambda}{dz} \, d\lambda = F = \frac{4}{3} \frac{1}{\kappa} \frac{dB}{dz} \tag{4.1}$$

we find the value of κ to use is the so-called Rosseland mean value $\bar{\kappa}_R$ which is given by (see also Appendix A.5)

$$\frac{1}{\bar{\kappa}_R} = \int_0^\infty \frac{1}{\kappa_\lambda} \frac{dB_\lambda}{dT} \, d\lambda \bigg/ \frac{dB}{dT} \tag{4.2}$$

where B_λ is the Planck function, namely

$$B_\lambda = \frac{2hc^2}{\lambda^5} \frac{1}{e^{hc/\lambda kT} - 1}$$

and (see also Volume 2)

$$B = \int_0^\infty B_\lambda \, d\lambda = \frac{\sigma}{\pi} T^4 \tag{4.3}$$

The Rosseland mean absorption coefficient is usually called the *opacity*.

In order to be correct we have to calculate the free-free absorption coefficients, the bound-free absorption coefficients and the line absorption coefficients for all λ and add those for each λ before taking the Rosseland mean of all these absorption coefficients. This will have to be done for all temperatures and pressures. The results of such computations have been published in the form of figures and tables. In Fig. 4.4 we show the opacities as a function of temperature for some densities, according to Cox and Tabor (1976). The opacities decrease with increasing temperature approximately as $\kappa \propto T^{-\delta}$ with $3.5 > \delta \geqslant 2.0$. They increase with increasing density approximately as $\kappa \propto \rho^{0.5}$. In the following discussion we will try to understand the general behavior of κ as a function of T. This will make it easier to understand the main features of stellar structure and their dependence on the stellar mass and chemical composition.

The bound-free absorption coefficient per hydrogen atom in the level with main quantum number n is given by

$$a_{\text{bf}}(n) = \frac{64\pi^4 m e^{10} Z'^4 g}{3\sqrt{3} \, ch^6 n^5 \nu^3} \tag{4.4}$$

Here g is the so-called Gaunt factor – a quantum mechanical correction factor to the absorption coefficients calculated otherwise in a classical way and making use of the correspondence principle (Gaunt was the first one to compute g). e is the elementary charge, m the mass of the electron, h Planck's constant and ν the frequency of the absorbed photon. Z' is the effective charge of the nucleus attracting the absorbing electron; Z' is 1 for the hydrogen atom. The absorption coefficient, except for the g factor, can be calculated classically by calculating the emission of an electron with a given impact parameter p which is accelerated in the Coulomb force field of the proton. A Fourier analysis and summation over all p gives the frequency dependence. (See for instance Finkelnburg and Peters, 1957.) It turns out that for hydrogen the quantum mechanical correction factor g for the bound-free transitions is never larger than a few per cent. For other atoms and ions we can use expression (4.2) as an approximation if we calculate the factor Z'^4 appropriately as the 'effective' charge of the nucleus, which is partially shielded by the remaining electrons. For atoms or ions other than hydrogen the correction factors g can be fairly large and in most cases are not even known. In stellar interiors, where hydrogen is completely ionized, only the heavy elements still have electrons able to absorb photons. The heavy elements are therefore very important for the absorption coefficient. Fortunately the 'hydrogen' like approximation given in equation (4.4) becomes more valid when the outer electrons are removed from the atoms. For the remaining highly ionized ions the energy level diagrams become more similar to the one for the hydrogen atom, except that the ionization energies are much larger, roughly proportional to Z'^2. Of the more abundant elements mainly iron has still a rather

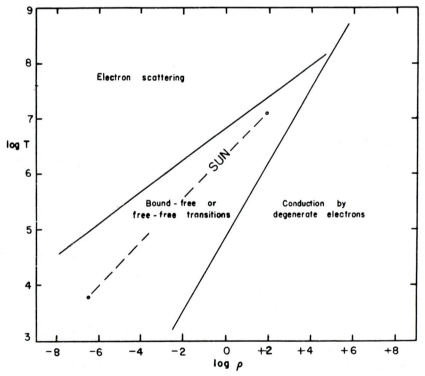

Fig. 4.3. In the temperature density diagram the regions are indicated for which the different absorption or scattering processes are important. Also shown are the high density regions for which heat conduction becomes important. The dashed line shows the approximate relation between temperature and density for the solar interior. Adapted from Schwarzschild (1958a).

complicated energy level diagram because even at temperatures of around a few million degrees it still has many bound electrons.

From equation (4.4) we see that the absorption coefficients for higher frequencies, i.e. for shorter wavelengths, generally decrease as v^{-3}. We also know that for higher temperatures the maximum of the Planck function shifts to shorter wavelengths. From Wien's law we know that $\lambda_{max} T = $ const. $= 0.289$ cm degree^{-1}, where λ_{max} is the wavelength for which the Planck function for the temperature T has a maximum, i.e. $v_{max} \propto T$. For the determination of the average $\bar{\kappa}_R$ the wavelengths with the largest flux are the most important. This means that $\bar{\kappa}_R$ is mainly determined by $\kappa(\lambda_{max}) \propto v_{max}^{-3} \propto T^{-3}$ according to (4.4). We may therefore expect a *decrease* in the opacity for higher temperatures as $\bar{\kappa}_R \propto T^{-3}$ due to the factor v^{-3} in κ. Of course, changes in Z' and in the number of ions which can contribute to the absorption also have to be considered. For the final bound-free absorption coefficient at a given frequency v we must sum up all the contributions from different energy levels for a given ion or atom and then sum up all the contributions from different ions or atoms. The final bound-free absorption coefficient is given by

$$\kappa_{v,cm} = \sum_{ions} \sum_{n} N_n a_{bf}(n) \qquad (4.5)$$

where N_n are the numbers of ions per cm^3 in a given quantum state with quantum number n, which can be calculated from the Saha equation and the Boltzmann formula. We have to total the contributions of all energy levels for all ions of all elements.

In order to obtain qualitative insight, for the summation over all the contributions we can make some simplifying assumptions. For increasing temperatures more and more electrons will be removed from the atoms. We therefore encounter higher and higher stages of ionization for increasing temperatures and $\bar{Z}'$ increases with increasing T, giving a slight increase in κ with T. The degree of ionization is described by the Saha equation which appears to introduce a factor of $T^{-3/2} n_e$ into the $\bar{\kappa}_R$ (n_e = electron density). However, the change from one state of ionization to the next one may not be so important.

The electrons come mainly from the ionization of hydrogen and helium. We get twice as many electrons per unit of mass from hydrogen as from helium. Therefore $n_e \propto \rho(2X + Y) = \rho(1 + X)$ since $X + Y \approx 1$, with X and Y being the mass fractions of the hydrogen and helium abundances.

Taking into account all these effects Kramers estimated that the Rosseland mean κ per gram of material for bound-free transitions depends on ρ and T as $\bar{\kappa}_R \propto \rho T^{-3.5}$. As pointed out above the main decrease in $\bar{\kappa}_R$ with increasing T is due to the shift of the radiative flux to shorter wavelengths where κ_λ is smaller, decreasing as ν^{-3}. Free-free transitions have to be considered in addition to the bound-free transitions. Their contribution has to be added at each frequency.

For temperatures of several million degrees in stellar interiors hydrogen and helium are completely ionized and also the heavy elements have lost most of their electrons. The free-free transitions then become very important. They are mainly due to transitions in the Coulomb field of the hydrogen and helium nuclei because they are so much more abundant than the heavy elements. The free-free absorption coefficient is therefore independent of the heavy element abundances. The dependence on ρ and T turns out, however, to be approximately the same as for bound-free transitions, namely $\bar{\kappa}_R \propto \rho \cdot T^{-3.5}$.

4.2   Electron scattering

We saw in Volume 2 that the scattering coefficient due to scattering on free electrons is independent of wavelength. The number of free electrons per gram also does not change with increasing temperature because almost all electrons come from hydrogen and helium, which are completely ionized for temperatures about 100 000 K. For mainly ionized material the electron scattering coefficient per gram does not therefore depend on density and temperature.

In Fig. 4.3 we illustrate the regions in the temperature density diagram where the different absorption or scattering processes are most important.

4.3   The line absorption coefficients

For a complete description of all the important absorption coefficients in stellar interiors we still have to include the contribution from the line absorption coefficients. Since the Rosseland mean absorption coefficient is a harmonic mean of all the κs, small wavelength bands with large values of κ_ν, as we see them in the spectral lines, are generally not expected to have a large influence on the $\bar{\kappa}$ unless they are so numerous that they cover a large fraction of all wavelengths. Pressure broadening of spectral lines becomes important for the high pressures in stellar interiors because it extends the wavelength coverage of the spectral lines. In the deep interior the highly ionized ions have energy levels which more and more resemble those of hydrogen, which means they have only a few energy levels with a relatively small number of line transitions. It turns out, however, that in the temperature range around a few hundred thousand to

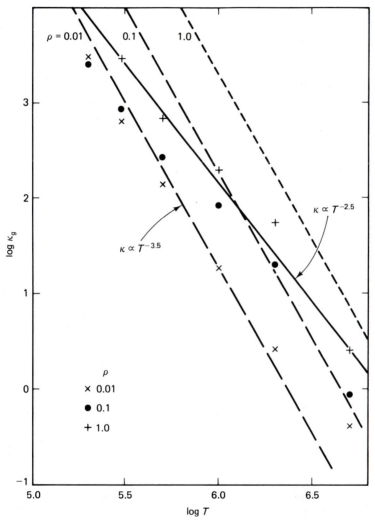

Fig. 4.4. The Rosseland mean absorption coefficient as a function of temperature T and density ρ according to Cox and Tabor (1976). A hydrogen abundance of 70 per cent by mass and a helium abundance of 28 per cent by mass was assumed with 2 per cent of heavy elements. For comparison some curves are shown with $\bar{\kappa}_g = C\rho T^{-3.5}$ for different values of ρ, with $C = 1.8 \times 10^{24}$ to match the point for $\rho = 0.01 \, \text{cm}^{-3}$ and $T = 10^6 \, \text{K}$. A line $\kappa_{gr} \propto \rho^{0.5} T^{-2.5}$ is also shown, which matches the numerical values better for $\rho \geqslant 0.1$.

a million degrees the heavy ions, like those of iron, still have a fairly large number of electrons in their outer shells creating a rather complicated energy level diagram with many line transitions. In this temperature region the line absorption coefficients are therefore still important. In Fig. 4.4 we compare the results of numerical calculations by Cox and Tabor (1976) with Kramers' estimate. An approximation $\bar{\kappa}_R = \kappa_g \propto \rho^{0.5} T^{-2.5}$ seems to be better than Kramers'.

5

Convective instability

5.1   General discussion

So far we have talked about energy transport by radiation only. We may also have energy transport by mass motions. If these occur hot material may rise to the top, where it cools and then falls down as cold material. The net energy transport is given by the difference of the upward transported energy and the amount which is transported back down. Such mass motions are also called convection. Our first question is: when and where do these mass motions exist, or in other words where do we find instability to convection? When will a gas bubble which is accidentally displaced upwards continue to move upwards and when will a gas bubble which is accidentally displaced downwards continue to move downwards? Due to the buoyancy force a volume of gas will be carried upwards if its density is lower than the density of the surroundings and it will fall downwards if its density is larger than that of the surroundings.

From our daily experience we know that convection occurs at places of large temperature gradients, for instance over a hot asphalt street in the sunshine in the summer, or over a radiator in the winter. The hot air over the hot asphalt, heated by the absorption of solar radiation, has a lower density than the overlying or surrounding air. As soon as the hot air starts rising by an infinitesimal amount, it gets into cooler and therefore higher density surroundings and keeps rising due to the buoyancy force like a hot air balloon in the cooler surrounding air. This always occurs if a rising gas bubble is hotter than its surroundings.

5.2   The Schwarzschild criterion for convective instability

When does a rising gas bubble obtain a higher temperature than the surrounding gas which is at rest? In Fig. 5.1 we demonstrate the situation. We have schematically plotted the temperature as a function of depth in the star. Suppose a gas bubble rises accidentally by a small distance in height. It gets into a layer with a lower gas pressure P_g and therefore it expands. Without any energy exchange with its surroundings it expands and cools adiabatically. If during this rise and adiabatic expansion the change in temperature is smaller than in the surroundings the gas bubble remains hotter than the surroundings. The expansion of the gas bubble, adjusting to the pressure of the surroundings, happens very fast, namely with the speed of sound. We can therefore assume that the pressure in the gas bubble and in the surroundings is the same (this is not quite true close to the boundaries of the convectively unstable zone) and therefore the higher temperature gas bubble will have a lower density than the surrounding gas. The buoyancy force will therefore accelerate it upwards. As seen in Fig. 5.1 this always occurs if the adiabatic change of temperature during expansion is smaller than the change of temperature with gas pressure in the surroundings. The condition for continuing motion, i.e. for convective instability, is then

$$\left(\frac{dT}{dz}\right)_{ad} < \left(\frac{dT}{dz}\right)_{sur} \tag{5.1}$$

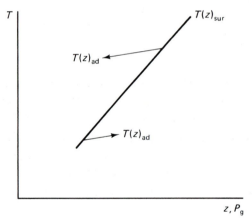

Fig. 5.1. The average relation $T(z)_{sur}$ or $T(P_g)_{sur}$ for a stellar atmosphere is shown schematically. Also shown is the $T(z)_{ad}$ relation for a rising or falling gas volume which for pressure equilibrium changes its temperature adiabatically when it rises and expands (cools) or falls and is compressed (with increasing T). If the temperature stratification $T(z)$ in the surrounding gas is steeper than for the adiabatic temperature stratification followed by the rising or falling gas, the rising gas will obtain a higher temperature and lower density than the surroundings and will be pushed up further by the buoyancy force, while the falling gas obtains a lower temperature and therefore higher density than the surroundings and keeps falling. The atmosphere is unstable to convection.

where ad stands for 'adiabatic' and sur stands for 'surroundings'. After replacing

$$\frac{dT}{dz} \quad \text{by} \quad \frac{dT}{dP_g}\frac{dP_g}{dz}$$

we can also say

$$\left(\frac{dT}{dP_g}\right)_{ad} < \left(\frac{dT}{dP_g}\right)_{sur} \tag{5.2}$$

because dP_g/dz is the same on both sides of the equation and is always positive. We easily see from Fig. 5.1 that under those conditions a falling gas bubble remains cooler than its surroundings and for equal pressure has therefore a higher density than the surroundings and keeps falling. For a layer in radiative equilibrium, convective instability then occurs if

$$\left(\frac{dT}{dP_g}\right)_{ad} < \left(\frac{dT}{dP_g}\right)_{radiative} \tag{5.3}$$

or

$$\mathbf{\nabla}_{ad} = \left(\frac{d\ln T}{d\ln P_g}\right)_{ad} < \left(\frac{d\ln T}{d\ln P_g}\right)_{rad} = \mathbf{\nabla}_{rad} \tag{5.4}$$

if we multiply equation (5.2) by P_g/T.

This is the **Schwarzschild criterion for convective instability**. In its derivation we have implicitly assumed that the mean molecular weight μ is the same in the rising bubble and in the surroundings. If this is not the case we have to take this into account when comparing the density in the bubble with that in the surrounding gas. If the deeper layers have a higher mean atomic weight, perhaps due to a higher abundance of helium, the density of the rising gas is increased and the convective instability decreased (see Fig. 5.2).

5.3 The adiabatic temperature gradient

In order to judge whether a given layer in a star is unstable against convection we have to compare the radiative equilibrium temperature gradient, given by equation (3.8), with the adiabatic temperature gradient. We therefore have to compute this adiabatic temperature gradient (see also Volume 2). By definition an adiabatic temperature and pressure change means a change with no heat exchange ΔQ with the surroundings, i.e.,

$$\Delta Q = 0 \tag{5.5}$$

with

$$dQ = dE + P_g\, dV \tag{5.6}$$

where dE is the internal energy, $dE = C_V\, dT$. Here C_V is the specific heat at constant volume (i.e. $dV = 0$). For an ideal gas we have

$$P_g V = R_g T \tag{5.7}$$

R_g is the gas constant, if we consider 1 mol of gas. V is then the volume for 1 mol, i.e. $V = \mu/\rho$. If the pressure is kept constant during heating the gas expands and has to do work against the outside pressure. This additional energy has to be supplied when heating the gas at constant pressure. The specific heat at constant pressure C_p per mol is therefore larger than C_V per mol. Differentiating equation (5.7) and making use of (5.6) (see also Volume 2) one finds

$$C_p - C_V = R_g \tag{5.8}$$

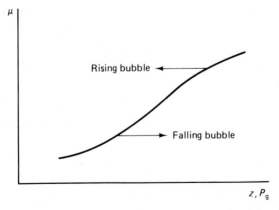

Fig. 5.2. μ is schematically shown as a function of depth in a hypothetical star. If in the interior the average atomic weight μ increases with depth a rising bubble carries with it a higher μ from the deeper layers. It therefore has a higher ρ than for the case with a depth independent μ. A falling bubble carries with it a lower μ which reduces its density. A positive $d\mu/dz$ makes convection more difficult.

For adiabatic changes for which $dQ = 0$ we then derive, introducing $\gamma = C_p/C_V$, that

$$\left(\frac{\rho}{\rho_0}\right)^\gamma = \frac{P_g}{P_{g0}} \quad \text{or} \quad P_g = P_{g0}\left(\frac{\rho}{\rho_0}\right)^\gamma \quad \text{and} \quad T \propto \frac{P}{\rho} \propto \rho^{\gamma-1} \qquad (5.9)$$

With this relation we can now easily determine ∇_{ad}. For an ideal gas we obtain

$$\left(\frac{d \ln T}{d \ln P_g}\right)_{ad} = \nabla_{ad} = 1 - \frac{1}{\gamma} = \frac{\gamma - 1}{\gamma} \qquad (5.10)$$

For a monatomic gas the internal energy per particle is generally $\frac{3}{2}kT$, if only the kinetic energy needs to be considered, and for a mol it is $E = \frac{3}{2}R_g T$. According to equation (5.8) $C_p = C_V + R_g$ and we have

$$C_p = \frac{5}{2}R_g \quad \text{and} \quad \gamma = C_p/C_V = \frac{5}{3} \qquad (5.11)$$

With (5.5) we therefore find

$$\nabla_{ad} = \frac{2}{3}/\frac{5}{3} = \frac{2}{5} = 0.4 \qquad (5.12)$$

for a monatomic gas for which $C_V = \frac{3}{2}R_g$.

Equation (5.10) tells us that ∇_{ad} becomes very small if γ approaches 1. Since

$$\gamma = \frac{C_p}{C_V} = \frac{C_V + R_g}{C_V} \quad \text{or} \quad \gamma = 1 + \frac{R_g}{C_V} \qquad (5.13)$$

it approaches 1 if C_V becomes very large. This happens if a large amount of energy is needed to heat the gas. In that case ∇_{ad} becomes small and convection can set in. We find such conditions for temperatures and pressures for which an abundant element like hydrogen or helium starts to ionize. A large amount of energy is needed to remove an electron from a hydrogen or helium atom; this energy is not available to increase the temperature. Therefore large amounts of energy are needed to heat the gas by one degree and at the same time supply the necessary ionization energy and ∇_{ad} becomes small.

Hydrogen starts to ionize for temperatures around 6000 to 7000 K. For temperatures higher than this we therefore find small values of ∇_{ad} until the temperatures become so high that all the hydrogen is ionized. At such temperatures ∇_{ad} increases again. However, ∇_{ad} becomes small again for temperatures around 20 000 to 50 000 K when helium ionizes once and twice. In Fig. 5.3 we show the contour lines for ∇_{ad} according to Unsöld (1948).

5.4   **Reasons for convective instabilities**

5.4.1   Convective instability due to a steep increase in the absorption coefficient

In order to find out where we have convection zones in stars we have to compare the radiative temperature gradient ∇_{rad} (see equation (3.8)) and the adiabatic temperature gradient ∇_{ad} and see under which conditions we find

$$\nabla_{rad} > \nabla_{ad} \tag{5.14}$$

This can be the case either because ∇_{rad} becomes very large or because ∇_{ad} becomes very small.

Let us first look at the radiative temperature gradient ∇_{rad} and see under which conditions it can become very large. From equation (3.8) and making use of

$$\frac{d \ln P_g}{dz} = \frac{\mu g}{R_g T} = \frac{1}{H}$$

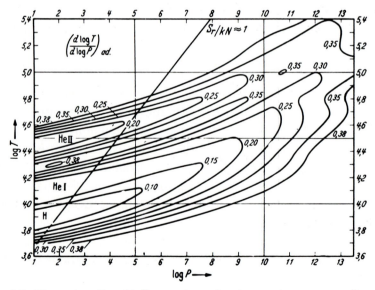

Fig. 5.3. The contour lines for ∇_{ad} are shown in a temperature pressure diagram. ∇_{ad} has a minimum for temperatures and pressures for which hydrogen or helium ionize. For the calculations the energy in the radiation field has been neglected. Above the line $S_r/kN = 1$ the energy in the radiation field should have been considered. In this region ∇_{ad} is actually smaller than calculated here. From Unsöld (1948), p. 231.

where H is the isothermal scale height (see Section 6.3.1), we find

$$\nabla_{\rm rad} = \frac{3\pi}{16} \frac{F\kappa_{\rm cm}H}{\sigma T^4} = \frac{3\pi}{16} \frac{F}{\sigma T^4} \kappa_{\rm g} \frac{P_g}{g} \tag{5.15}$$

with $\kappa_{\rm cm} = \kappa_{\rm g}\rho$ (see Chapter 3). Of course, for radiative equilibrium $F = F_{\rm r}$.

The radiative gradient becomes very large if either the flux F becomes very large for a given T or if $\kappa_{\rm g}P_g$ becomes very large. In Volume 2 we saw that generally at a given optical depth $P_g \propto \kappa_{\rm g}^{-1}$; therefore for depth independent $\kappa_{\rm g}$ the product $P_g\kappa_{\rm g}$ is not expected to vary much. Suppose $\kappa_{\rm g}$ is small at the surface. Then at the optical depth $\tau = 1$ for instance P_g will be large because $\tau = 1$ corresponds to a rather deep layer in the atmosphere. If on the other hand for $\tau > 1$ the $\kappa_{\rm g}$ increases steeply, then we still have a large P_g because of the small $\kappa_{\rm g}$ close to the surface. With a large $\kappa_{\rm g}$ for $\tau \geqslant 1$ the product $P_g\kappa_{\rm g}$ can then become very large. **It is the depth dependence of $\kappa_{\rm g}$ which is important for the value of $\nabla_{\rm rad}$.** This can be clearly seen if we now calculate $\nabla_{\rm rad}$ for a special case in which we approximate the depth dependence of $\kappa_{\rm g}$ by (see also Volume 2)

$$\kappa_{\rm g} = \kappa_0 P_g^b(z) \tag{5.16}$$

A steep increase in $\kappa_{\rm g}$ with depth is found for large values of b.

We calculate the pressure gradient from the hydrostatic equilibrium equation

$$\frac{{\rm d}P_g}{\kappa_{\rm cm}\,{\rm d}z} = \frac{{\rm d}P_g}{{\rm d}\tau} = \frac{g\rho}{\kappa_{\rm cm}} = \frac{g}{\kappa_{\rm g}} = \frac{g}{\kappa_0} \frac{1}{P_g^b}$$

or separating the variables

$$P_g^b\,{\rm d}P_g = \frac{g}{\kappa_0}\,{\rm d}\tau \tag{5.17}$$

After integration between $\tau = 0$ and τ, for which $P_g = 0$ and $P_g = P_g(\tau)$ respectively, we obtain

$$\frac{1}{b+1}P_g^{b+1}(\tau) = \tau\frac{g}{\kappa_0} \quad \text{or} \quad P_g^{b+1}(\tau) = (b+1)\tau\frac{g}{\kappa_0} \tag{5.18}$$

Equations (5.17) and (5.18) yield

$$\frac{{\rm d}\ln P_g}{{\rm d}\tau} = \frac{1}{P_g}\frac{{\rm d}P_g}{{\rm d}\tau} = \frac{g}{\kappa_0}\frac{1}{P_g^{b+1}} = \frac{g}{\kappa_0(b+1)}\frac{\kappa_0}{\tau g} = \frac{1}{(b+1)\tau} \tag{5.19}$$

We now calculate ∇_{rad}. We had derived previously (equation (3.6)) that

$$\frac{4}{3}\frac{dB}{d\tau} = F_r$$

The integration of this relation from $\tau = 0$ to τ yields for $F_r = F = \text{constant}$, i.e. for radiative equilibrium in the plane parallel case,

$$B(\tau) = \tfrac{3}{4}F\tau + B_0 \approx \tfrac{3}{4}F\tau \qquad (5.20)$$

for large τ when $B_0 = B(\tau = 0) \ll B(\tau)$.

With

$$B = \frac{\sigma}{\pi}T^4 \qquad (5.21)$$

we find

$$\left(\frac{dB}{d\tau}\right)_{rad} = \frac{dB}{dT}\left(\frac{dT}{d\tau}\right)_{rad} = \frac{4\sigma}{\pi}\frac{T^4}{T}\left(\frac{dT}{d\tau}\right)_{rad}$$

$$= 4B\left(\frac{d\ln T}{d\tau}\right)_{rad} = 3F\tau\left(\frac{d\ln T}{d\tau}\right)_{rad} \qquad (5.22)$$

using equation (5.20), or

$$\left(\frac{d\ln T}{d\tau}\right)_{rad} = \left(\frac{dB}{d\tau}\right)_{rad}\frac{1}{3F\tau} = \frac{1}{4\tau} \qquad (5.23)$$

As

$$\nabla_{rad} = \left(\frac{d\ln T}{d\ln P_g}\right)_{rad} = \left(\frac{d\ln T}{d\tau}\right)_{rad}\bigg/\frac{d\ln P_g}{d\tau} \qquad (5.24)$$

we now have to divide equation (5.23) by $(d\ln P_g/d\tau)$ from the hydrostatic equation (see equation (5.19)), and obtain for $\tau \gg 1$

$$\nabla_{rad} = \frac{1}{4\tau}\bigg/\frac{1}{(b+1)\tau} = \frac{b+1}{4} \qquad (5.25)$$

Only for $b \geqslant 0.6$ do we find $\nabla_{rad} \geqslant 0.40$. If $\nabla_{ad} = 0.40$, i.e. if $\gamma = \tfrac{5}{3}$, we will find convective instability for $b > 0.6$, which sets a lower limit on the *increase* in κ for convective instability but not on the absolute value of κ_g determined by κ_0. For the surface layers ∇_{rad} does not depend on κ_0. For constant κ_g we find $\nabla_{rad} = 0.25$ no matter how large κ_0 is.

For constant κ_g we need $\nabla_{ad} < 0.25$ for convective instability.

In the temperature region where the hydrogen starts to ionize, i.e. around 6000 to 7000 K, the absorption coefficient increases very steeply for two reasons:

1. In the surface layers of cool stars the H^- ions do most of the continuous absorption. Due to the beginning of hydrogen ionization the number of free electrons increases and many more H^- ions are formed, therefore the H^- absorption coefficient increases steeply.

2. The temperature is high enough to excite the higher energy levels in the hydrogen atoms, which means we find a large number of electrons in the second and third energy levels, the Balmer and Paschen levels, of the hydrogen atom, which can then contribute to the continuous absorption coefficient for $\lambda < 8200$ Å. This gives a very steep increase in the hydrogen absorption coefficient in the wavelength region in which we find most of the radiative flux. In this upper part of the hydrogen ionization zone therefore κ_g increases very steeply and ∇_{rad} can exceed 1000.

5.4.2 Convective instability due to large energy flux F

We also find convective instability when the energy flux F becomes very large. This can happen in the interiors of massive stars. In these stars the central temperatures are so high that the CNO cycle (see section 8.5) is the main energy source. We know that the CNO cycle is very temperature sensitive; the energy generation is approximately proportional to T^{16}. This means that the energy generation is strongly concentrated towards the center of the stars where the temperatures are highest. The total luminosity is generated in a very small sphere with radius r_1 around the center. If we now calculate the energy flux F from $\pi F = L/4\pi r_1^2$ we find a very large value for F because r_1 is so small. In the center of massive stars we therefore also find convective instability due to the large value for F. As we pointed out this is due to the strong temperature dependence of the CNO cycle. We therefore find these core convection zones only in stars where the CNO cycle is important for the energy generation, that is in hot stars.

5.4.3 Convective instability due to small values of the adiabatic gradient ∇_{ad}

We discussed above that the adiabatic temperature gradient becomes very small in the temperature regions where an abundant element like hydrogen or helium ionizes, i.e. in the hydrogen and helium ionization zones. In Section 5.4.1 we saw that these are also the regions where the continuous κ increases steeply and where we therefore find convective instability. In the hydrogen ionization region the ∇_{ad} becomes small and ∇_{rad} becomes very large which enhances the convective instability (see Fig. 5.3). The small values of ∇_{ad} extend the convection zones but they do not create new ones in this temperature region. The instability ceases when the hydrogen and helium ionizations are essentially complete.

5.4.4   *Convective instability due to molecular dissociation*

For completeness we should mention that small values of ∇_{ad} are also found in stellar atmospheric regions where the hydrogen molecule starts to dissociate. In these regions a large amount of energy is needed to increase the temperature by 1 degree because for increasing temperatures more molecules dissociate and use energy in this way which is then not available for an increase in kinetic energy, i.e. for an increase in temperature. The specific heat C_V is then very large and ∇_{ad} is small, leading to a high atmospheric convection zone in cool stars.

5.4.5   *Summary*

In hot stars, $T_{eff} > 9000$ K, we find core convection zones but envelopes in radiative equilibrium, because hydrogen is already ionized in the surface layers. Helium ionization still leads to convective instability but with negligible energy transport. The densities in this ionization zone are too low.

In cool stars, $T_{eff} < 7600$ K, we find thin surface layers in radiative equilibrium, below which so-called hydrogen convection zones are found which are very thin if $T_{eff} > 7000$ K.

For $T_{eff} < 7000$ K, hydrogen convection zones are found for layers with $T \gtrsim 6000$ K with adjacent helium convection zones. Below this outer convection zone we again find radiative equilibrium zones down to the center of the star, except for very cool stars for which the hydrogen and helium convection zones are very extended and may for main sequence stars perhaps reach down to the center of the stars.

For very cool stars, i.e. M stars, we may also find surface convection zones due to the dissociation of hydrogen molecules. These unstable zones are separated from the hydrogen convection zones by regions in radiative equilibrium. Because of the low densities in these layers near the surface, convective energy transport is not expected to be important in these cool star surface convection zones.

6

Theory of convective energy transport

6.1 Basic equations for convective energy transport

If we find convective instability in a given layer of a star we also have to consider energy transport by convection. If part of the energy is transported by convection, less energy needs to be transported by radiation; πF_r is then decreased and according to equation (3.8) the temperature gradient needs to be less steep than in radiative equilibrium. The total energy transport of course has to be the same as in the case without convection, otherwise we could not have thermal equilibrium. We therefore require that

$$\pi F_r + \pi F_c = \pi F = \sigma T_{eff}^4 \frac{R^2}{r^2} \qquad (6.1)$$

where R is the stellar radius and r is the radius of the layer for which we want to calculate the flux. F_c is the convective flux.

In order to calculate the temperature gradient necessary to transport the required amount of radiative flux F_r we must determine the amount of convective energy flux F_c.

The net convective energy transport is given by the difference of the energy transported upwards and the amount transported back down again. We look at two columns in a given layer of the star (see Fig. 6.1): in one column the gas is moving upwards, in the other the gas is moving downwards. All variables referring to the column with upward (downward) moving gas have the subscript u (d). The cross-sections for the columns are then σ_u and σ_d respectively. The velocities are v, the densities ρ, and the temperatures T. We now consider the energy flux through a cross-section of the two columns. Each second a column of length v moves through this cross section. The amount of material moving upwards is given by $\rho_u v_u \sigma_u$ and the amount moving downwards by $\rho_d v_d \sigma_d$. In the upward moving column the heat content per gram is given by $e_u = c_p T_u$, and for the downward moving column $e_d = c_p T_d$, where we have assumed equal pressure in the columns. Here c_p is the specific heat for constant pressure per gram of material. For the net energy transport through $\sigma_u + \sigma_d$ we find if $\sigma_u = \sigma_d = 1 \text{ cm}^2$

$$\rho_u v_u c_p T_u + \tfrac{1}{2}\rho_u v_u v_u^2 - (\rho_d v_d c_p T_d + \tfrac{1}{2}\rho_d v_d v_d^2) = 2\pi F_c \qquad (6.2)$$

where πF_c is the convective energy transport through 1 cm^2. (We are actually looking at a cross-section of 2 cm^2 here, hence the factor 2 on the right-hand side.)

We now have to consider the mass transport. There obviously cannot be any net mass transport in any one direction, because otherwise the star would dissolve or collect all the mass in the interior which would violate hydrostatic equilibrium. We must require that the net mass transport is zero, which means

$$\rho_u v_u = \rho_d v_d = \rho v \quad \text{which for} \quad \rho_u \approx \rho_d \quad \text{yields} \quad v_u = v_d \qquad (6.3)$$

provided that the cross-sections for the upward and downward moving material are the same, otherwise $\rho_u v_u \sigma_u = \rho_d v_d \sigma_d$. From equation (6.3) we find that the net transport of kinetic energy is zero. There is as much kinetic energy carried upwards as is carried back down. (If $\sigma_u \neq \sigma_d$ the velocities may be different and the kinetic energy transport may not be zero. In any case the kinetic energies are subtracted from each other. Only a small fraction contributes to the net energy flux.)

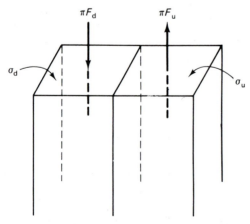

Fig. 6.1. Through a column with cross-section σ_u a stream of gas is flowing upwards, carrying an energy flux $\sigma_u \pi F_u$. Through another column with cross-section σ_d gas is flowing back down, carrying an energy flux $\sigma_d \pi F_d$.

Making use of equation (6.3) we derive for the net convective energy flux (see also Volume 2) the expression

$$\pi F_{\mathrm{c}} = \tfrac{1}{2}\rho v c_p (T_{\mathrm{u}} - T_{\mathrm{d}}) \approx \rho v c_p \Delta T \tag{6.4}$$

which gives us the energy transport per second through 1 cm^2 by means of mass motion, generally called convective energy transport. $\Delta T = T_{\mathrm{u}} - T_{\mathrm{s}} = T_{\mathrm{s}} - T_{\mathrm{d}}$, such that $T_{\mathrm{u}} - T_{\mathrm{d}} = 2\Delta T$ (see Fig. 6.2). T_{s} is the average temperature of the surroundings.

Clearly the assumption of equal density and equal pressure in the two columns, while permitting temperature differences, is somewhat inconsistent, but introducing $\Delta \rho$ would introduce second-order terms in the heat transport, which we neglect.

For unequal densities the energy transport of kinetic energy would not exactly cancel; however, the remaining difference would generally be very small in comparison with the heat transport term except possibly close to the boundaries of the convection zones, where the gas pressures in the two columns and the cross-sections σ may be different. In equation (6.4) we find $F_{\mathrm{c}} \propto T_{\mathrm{u}} - T_{\mathrm{d}} \approx 2\Delta T$. This means the positive energy transport upwards and the negative energy transport downwards add up in the net heat

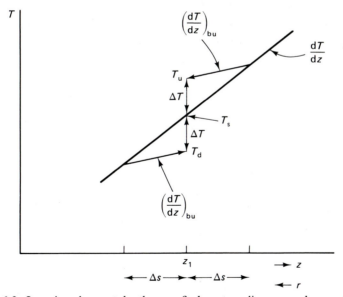

Fig. 6.2. In a given layer at depth z_1 we find gas traveling upwards as well as downwards with a temperature T_{u} or T_{d} respectively. T_{s} is the average temperature at depth z_1. For equal traveling distances Δz for the gas the difference $T_{\mathrm{u}} - T_{\mathrm{d}} = 2\Delta T$, with $\Delta T = T_{\mathrm{u}} - T_{\mathrm{s}}$ and also $\Delta T = T_{\mathrm{s}} - T_{\mathrm{d}}$. The subscript bu stands for gas bubble.

energy transport upwards. The negative energy transport downwards has the same effect as the positive energy transport upwards.

The convective flux is then generally given by

$$\pi F_c = \rho v c_p \Delta T \tag{6.5}$$

The question remains how to calculate ΔT and the velocity v. If we look at all the different rising and falling columns in a given layer at depth z_1 we find many different values for ΔT and v, depending on how far the gas has already traveled (see Fig. 6.3). What we are interested in is the overall energy transport, which means we need to know the average ΔT and v at a given depth which depend on the average traveling distance which these gas columns or bubbles, crossing a given horizontal layer, have been traveling. We shall call the average traveling distance for a gas bubble prior to mixing with the surroundings l. This distance is frequently called the mixing length when it is assumed that this distance is determined by the mixing of the rising and falling gas.

6.2   Mixing length theory of convective energy transport

In order to determine ΔT and v we now have to follow a given volume of gas, which we call a bubble, on its way upwards or downwards along the path s. We choose an upwards moving bubble which experiences

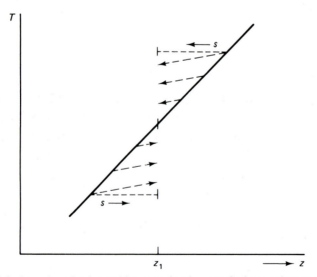

Fig. 6.3. In a given horizontal layer at depth z_1 we find gas columns traveling upwards or downwards which have originated at different layers and have traveled different distances s. They have different values of ΔT and v.

a temperature gradient $(dT/dz)_{bu}$, while z stands for depth. From Fig. 6.3 we see that $\Delta T(s)$ is determined by the difference between $(dT/dz)_{bu}$ and the average temperature gradient in the surroundings dT/dz. If s describes the distance which the bubble has traveled we find

$$\Delta T(s) = \left[\frac{dT}{dz} - \left(\frac{dT}{dz} \right)_{bu} \right] s \qquad (6.6)$$

As soon as the bubble develops a ΔT with respect to its surroundings there will be some energy exchange which decreases ΔT. The actual $(dT/dz)_{bu}$ will therefore be larger than the adiabatic gradient. We have

$$\Delta T(s) < \left[\frac{dT}{dz} - \left(\frac{dT}{dz} \right)_{ad} \right] s \qquad (6.7)$$

Due to the positive ΔT the rising bubble experiences a buoyancy force f_b. It is equal to the difference in gravitational force working on the bubble and on the surroundings. The force per cm^3 is

$$f_b = \Delta \rho g \quad \text{and the acceleration } a \text{ is} \quad a = \frac{\Delta \rho}{\rho} g = \frac{f_b}{\rho} \qquad (6.8)$$

Here g is the gravitational acceleration, pointing in the direction of z.

If we assume pressure equilibrium between the bubble and the surrounding gas we find for the difference in density between the bubble and the surrounding gas

$$\frac{\Delta \rho}{\rho} = -\frac{\Delta T}{T} + \frac{\Delta P_g}{P_g} = -\frac{\Delta T}{T} \qquad (6.9)$$

for $\Delta P_g = 0$ and assuming that the mean molecular weight is the same in the bubble as in the surrounding gas. Even for the same chemical composition this may not be the case if the degree of ionization changes due to the temperature and density change. Taking this into account we find

$$\frac{\Delta \rho}{\rho} = -\frac{\Delta T}{T} + \frac{\Delta \mu}{\mu} = -\Delta \ln T \left(1 - \frac{d \ln \mu}{d \ln T} \right) = -\Delta \ln T \cdot Q \qquad (6.10)$$

where

$$Q = 1 - \frac{d \ln \mu}{d \ln T}$$

Considering for simplicity the case of constant μ, i.e. $Q = 1$, we can express the buoyancy force as

$$f_b = -\frac{\Delta T}{T} \rho g = \rho \frac{dv}{dt} \qquad (6.11)$$

With the kinetic energy $E_{kin} = \frac{1}{2}\rho v^2$ being equal to $\int$ force ds we find

$$\frac{1}{2} v^2 = \int_0^s \frac{\Delta T}{T} g \, ds \quad \text{and} \quad v^2 = 2 \int_0^s \frac{\Delta T}{T} g \, ds \qquad (6.12)$$

According to equation (6.7) we have then

$$v^2(s) = 2g \int_0^s \frac{\left[\dfrac{dT}{dz} - \left(\dfrac{dT}{dz}\right)_{bu}\right]}{T} s \, ds = \frac{2g\left[\dfrac{dT}{dz} - \left(\dfrac{dT}{dz}\right)_{bu}\right]}{T} \frac{s^2}{2} \qquad (6.13)$$

This yields

$$v(s) = \left[\frac{g}{T}\left(\frac{dT}{dz} - \left(\frac{dT}{dz}\right)_{bu}\right)\right]^{1/2} s \qquad (6.14)$$

If we now compute the total flux through a given horizontal layer, we have to average over all bubbles crossing this layer. They have different travel distances s. With the average total travel distance for the bubbles being l, the average travel distance for the gas crossing at a given horizontal layer is $l/2$ (see Fig. 6.3) and the expression for the convective energy transport finally becomes

$$\pi F_c = \rho c_p T \frac{\bar{\Delta T}}{T} \bar{v}$$

$$= \rho c_p T \frac{\Delta T}{T} \left(\frac{g}{T}\right)^{1/2} \left[\frac{dT}{dz} - \left(\frac{dT}{dz}\right)_{bu}\right]^{1/2} \frac{l}{2}$$

$$= \rho c_p T \sqrt{g} \frac{\left[\dfrac{dT}{dz} - \left(\dfrac{dT}{dz}\right)_{bu}\right]^{3/2}}{T^{3/2}} \left(\frac{l}{2}\right)^2 \qquad (6.15)$$

If we take into account the possibility of changing ionization we find

$$\pi F_c = \rho c_p T Q \sqrt{g} \frac{\left[\dfrac{dT}{dz} - \left(\dfrac{dT}{dz}\right)_{bu}\right]^{3/2}}{T^{3/2}} \left(\frac{l}{2}\right)^2 \qquad (6.16)$$

Here we have used

$$\frac{\bar{\Delta}T}{T} = \frac{\overline{\left[\dfrac{\mathrm{d}T}{\mathrm{d}z} - \left(\dfrac{\mathrm{d}T}{\mathrm{d}z}\right)_{\mathrm{bu}}\right]}}{T}\frac{l}{2} \quad \text{and} \quad \bar{v} = \left\{\frac{g}{T}\left[\frac{\mathrm{d}T}{\mathrm{d}z} - \left(\frac{\mathrm{d}T}{\mathrm{d}z}\right)_{\mathrm{bu}}\right]\right\}^{1/2}\frac{l}{2} \quad (6.17)$$

With increasing $\left[\dfrac{\mathrm{d}T}{\mathrm{d}z} - \left(\dfrac{\mathrm{d}T}{\mathrm{d}z}\right)_{\mathrm{bu}}\right]$ the $\dfrac{\bar{\Delta}T}{T}$ and πF_{c} both increase.

6.3   Choice of characteristic travel length *l*

6.3.1   *The pressure scale height*

We expect the characteristic travel length *l* to be comparable to the pressure scale height H. It is determined by the hydrostatic equation

$$\frac{\mathrm{d}P}{\mathrm{d}r} = -g\rho \qquad (6.18)$$

where

$$\rho = \frac{P_g\mu}{R_g T} \qquad (6.19)$$

and g is the gravitational acceleration.

Assuming again $P \sim P_g$ and inserting (6.19) and (6.18) we find after dividing by P_g

$$\frac{1}{P_g}\frac{\mathrm{d}P_g}{\mathrm{d}r} = \frac{\mathrm{d}\ln P_g}{\mathrm{d}r} = \frac{-g\mu}{R_g T} \qquad (6.20)$$

By integrating this differential equation we find the pressure stratification:

$$\int_{P_{g0}}^{P_{g1}} \mathrm{d}\ln P_g = -\int_{r_0}^{r_1} \frac{g\mu}{R_g T}\,\mathrm{d}r \qquad (6.21)$$

where P_{g0} is the gas pressure at $r = r_0$ and P_{g1} is the gas pressure at radius r_1. In order to integrate this equation we need to know $T(r)$ and $g(r)$. For the moment we assume for simplicity a layer in which $T = \text{const.} = \bar{T}$ and

$g = \text{const.} = \bar{g}$, i.e. we calculate the gas pressure for an *isothermal* atmosphere. The integrand on the right-hand side of equation (6.21) is then constant and we find

$$\ln P_{g1} - \ln P_{g0} = \ln\left(\frac{P_{g1}}{P_{g0}}\right) = -\frac{\mu g}{R_g T}(r_1 - r_0) \qquad (6.22)$$

Taking the exponential for both sides we derive

$$P_{g1} = P_{g0} \exp\left[-\frac{\mu g}{R_g T}(r_1 - r_0)\right] = P_{g0} \exp\left[-\frac{r_1 - r_0}{H}\right] \qquad (6.23)$$

with

$$H = \frac{R_g T}{\mu g} \qquad (6.24)$$

For $r_1 - r_0 = H$ the pressure P_g changes by a factor of e. H is therefore called the isothermal pressure scale height. We can easily verify that this is also the isothermal density scale height. With temperature gradients the pressure and density scale heights will be different, but H still gives a good approximation if we use an average T.

6.3.2   *Relation between pressure scale height and characteristic length l*

If originally in a convective layer the cross-sections of the rising and falling gas columns were equal then after one pressure scale height H the rising gas must have expanded by a factor of e, if it remains in pressure equilibrium with the surroundings. This means after a distance comparable to H there is no room left for the falling gas (see Fig. 6.4). On the other hand we must require that through every horizontal layer the same amount of material must fall down as is moving upwards. There is only one solution to the problem, namely that not all the material starting to rise can keep going; part of it must be dragged down again by the falling material. In any given layer a large fraction of the falling material must have been taken out of the rising columns only a short time earlier. We therefore estimate that l cannot be much larger than H. In practical applications generally only the isothermal scale height is used. In this approximation the density and the pressure scale heights are of course identical. The actual value of l/H to be used can be determined by the comparison of real stars with stellar models calculated by assuming different values for the characteristic length. The larger the characteristic length, the more efficient is the convective energy transport, as seen from equation (6.15).

6.4   Energy exchange between rising or falling gas and surroundings

In order to calculate the convective energy transport we still have to determine the temperature gradient for the rising or falling bubbles; this is steeper than the adiabatic one because of the energy exchange with the surroundings. This energy exchange depends on the size of the gas bubbles. The larger the bubble, the more difficult it is for the photons to escape. We have to use an average size which we can only guess. Observations of solar granulation (see Volume 2) suggest sizes which are of the order of one pressure scale height. For lack of better knowledge we use this value, or the value of l.

Generally we can say that the ratio of the energy gain of the rising gas, with respect to the surroundings, to the energy loss (to the surroundings) is given by

$$\Gamma = \frac{\text{energy gain during lifetime}}{\text{energy loss during lifetime}} = \frac{c_p \rho \Delta T_{\max}}{c_p \rho \Delta T'} = \frac{\dfrac{\mathrm{d}T}{\mathrm{d}z} - \left(\dfrac{\mathrm{d}T}{\mathrm{d}z}\right)_{\mathrm{bu}}}{\left(\dfrac{\mathrm{d}T}{\mathrm{d}z}\right)_{\mathrm{bu}} - \left(\dfrac{\mathrm{d}T}{\mathrm{d}z}\right)_{\mathrm{ad}}} = \frac{E_{\mathrm{gain}}}{E_{\mathrm{loss}}}$$

$$(6.25)$$

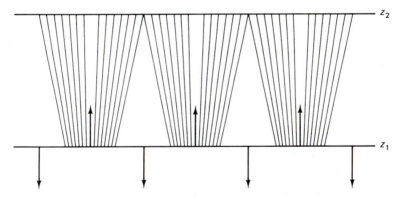

Fig. 6.4. In a layer at depth z_1 half the area is covered by rising bubbles and half by falling gas. The rising bubbles expand. They cover larger and larger areas. If in the layer at depth z_2 their cross-section has expanded by a factor of 2, no areas would be left for the falling gas. The volume of the rising gas expands by a factor of e over one scale height. The cross-section expands by a factor of 2 over a distance of about one density scale height. Most of the upwards traveling gas has to be diverted into the downward stream over this distance. The characteristic traveling distance l is therefore expected to be of the order of one scale height.

where $\Delta T_{\max}$ is the maximum temperature difference actually achieved between bubble and surroundings after it has traveled the distance l, and $\Delta T'$ is the change in temperature due to the energy exchange (see Fig. 6.5).

In order to calculate Γ we have to compute the energy loss which is due mainly to radiative energy exchange. We call $F_{(radial)}$ the radiative flux that is streaming out of the surface of the bubble in all directions. For large bubbles the energy loss per second is then $\pi F_{(radial)} \cdot$ surface area of the bubble and

$$E_{\text{loss}} \propto \frac{\mathrm{d}B}{\mathrm{d}\tau_r} \cdot \text{surface area of the bubble} \cdot \frac{l}{v} \tag{6.26}$$

τ_r is the radial optical depth through the bubble in any direction, $\tau_r = \kappa_{cm} l$. l/v is the lifetime of the bubble.

The total energy gain of the bubble is given by

$$E_{\text{gain}} = c_p \rho \Delta T_{\max} \cdot \text{volume} \tag{6.27}$$

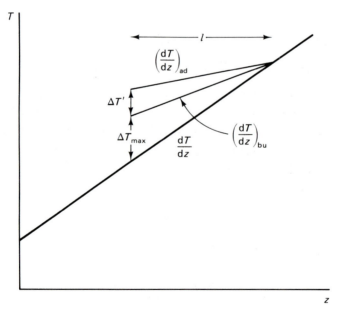

Fig. 6.5. The temperature stratification in an atmosphere with the temperature gradient $\mathrm{d}T/\mathrm{d}z$. The stratification with the adiabatic temperature gradient $(\mathrm{d}T/\mathrm{d}z)_{ad}$ and the actual temperature change of the bubble $(\mathrm{d}T/\mathrm{d}z)_{bu}$. Without energy exchange the bubble would follow the adiabatic temperature change. Due to energy exchange its temperature is decreased by $\Delta T'$ and the energy content per cm^3 reduced by $c_p \rho \Delta T'$.

We thus find

$$\Gamma = \text{const.} \frac{c_p \cdot \rho \cdot T \Delta T_{\max} \cdot \kappa_{cm} \cdot l \cdot \text{volume} \cdot v}{\sigma T^4 \cdot \Delta T \cdot \text{surface} \cdot l} = \Gamma_0 \cdot v \qquad (6.28)$$

More energy is lost for smaller velocities, i.e. longer lifetimes.

The value of the constant depends on the geometry of the bubble and on the temperature stratification in the bubble. Böhm-Vitense (1958) estimated Γ_0 to be

$$\Gamma_0 = \frac{1}{24} \frac{c_p \rho T \kappa_{cm} l}{\sigma T^4} \qquad (6.29)$$

It is a function of T and P_g and can be evaluated locally.

Equations (6.1), (6.4), (6.17), (6.18) and (6.25) constitute a set of equations from which the five unknowns dT/dz, $(dT/dz)_{bu}$, v, F_r, and F_c can be determined. Methods for the solution of these equations have been discussed by Böhm-Vitense (1953) and by Kippenhahn, Weigert and Hofmeister (1967). Here we will not go into details, but instead discuss only the results for some limiting cases: when the energy exchange is very small, i.e. $\Gamma \ll 1$, and when the energy exchange is very large, i.e. $\Gamma \gg 1$.

6.5   Temperature gradient with convection

6.5.1   *General expressions for ∇ and ∇_{bu}*

We defined the radiative temperature gradient $(dT/dz)_{rad}$ as the gradient which has to be present for radiative equilibrium, i.e. for $F_r = F$. This temperature gradient is given by equation (3.8) if we replace F_r by $F = \sigma T_{eff}^4$ in the case of a plane parallel atmosphere and $d/dz = -d/dr$.

Making use of the hydrostatic equation

$$\frac{d \ln P}{dz} = \frac{1}{H} \qquad (6.30)$$

we found for ∇_{rad} (see equation (5.21))

$$\nabla_{rad} = \frac{3}{16} \left(\frac{T_{eff}}{T} \right)^4 \bar{\kappa}_{cm} H = \frac{3}{16} \left(\frac{T_{eff}}{T} \right)^4 \frac{\bar{\kappa}_g P_g}{g} \qquad (6.31)$$

In case of spherical symmetry we have $\pi F_r = \sigma T_{eff}^4 (R^2/r^2)$ and the gravitational acceleration is $g = g_0(R^2/r^2)$ as long as the mass $M_r \approx M$ and where g_0 is the surface gravity. The factors r^2 cancel and we still find the same expression for ∇_{rad}.

For ∇_{ad} we found (see equation (5.10)) for constant γ

$$\nabla_{ad} = \left(\frac{d \ln T}{d \ln P_g}\right)_{ad} = (\gamma - 1)/\gamma \qquad (6.32)$$

For $C_p/C_V = \frac{5}{3}\gamma$ we find $\nabla_{ad} = 0.40$.

∇_{bu} is the corresponding logarithmic temperature gradient for the rising bubble. (Notice that the ∇ are all positive providing P_g increases inwards.)

We can rewrite the equations for $\Delta T, \bar{v}$ and the temperature gradient in terms of these logarithmic gradients.

The condition of thermal equilibrium requires that the sum of all energy fluxes has to be independent of depth. In the case of radiative and convective energy flux this means in the plane parallel case.

$$F = F_r + F_c = \text{const.} = \frac{\sigma}{\pi} T_{eff}^4 \qquad (6.33)$$

or

$$F_r = F - F_c = \frac{\sigma}{\pi} T_{eff}^4 - F_c \qquad (6.34)$$

The relation between the temperature gradient and the radiative flux remains unchanged (see Chapter 3, equation (3.6))

$$F_r = \frac{4}{3}\frac{dB}{d\bar{\tau}} = \frac{\sigma}{\pi}\frac{16}{3}T^4 \frac{1}{\bar{\kappa}_{cm}}\frac{d \ln T}{dz} \qquad (6.35)$$

Division by $d \ln P_g/dz = 1/H$ and solving for $d \ln T/d \ln P_g$ yields

$$\frac{d \ln T}{d \ln P_g} = \nabla = \pi F_r \frac{3}{16}\frac{\bar{\kappa}_{cm}H}{\sigma T^4} \qquad (6.36)$$

The temperature gradient ∇ is proportional to the radiative flux F_r. With convection $\pi F_r < \pi F$ and therefore the temperature gradient becomes smaller than for radiative equilibrium, i.e. $\nabla < \nabla_{rad}$. With $F_r = F - F_c = (\sigma/\pi)T_{eff}^4 - F_c$ we find

$$\nabla = (\sigma T_{eff}^4 - \pi F_c)\frac{3\bar{\kappa}_{cm}H}{16\sigma T^4} \qquad (6.37)$$

from which we derive

$$\nabla = \frac{\nabla_{rad} + \beta\nabla_{bu}}{1 + \beta} \qquad (6.38)$$

after replacing πF_c by the expression given in equation (6.5). Here $\beta \approx$ const. $\cdot \Gamma$. The constant depends on the constant in Γ (equation 6.28)). After some algebra one finds with $\beta = \frac{3}{2}\Gamma$

$$\nabla_{\text{bu}} = \frac{\nabla_{\text{rad}} + \Gamma(1 + \frac{3}{2}\Gamma)\nabla_{\text{ad}}}{1 + \Gamma(1 + \frac{3}{2}\Gamma)} \qquad (6.39)$$

These equations contain the unknown velocity because $\Gamma = \Gamma_0 v$ and $\beta \propto \Gamma$. We therefore cannot solve these equations straightforwardly, because v in turn depends on ∇ and ∇_{bu}.

For the special cases of very large or very small energy exchange between the bubbles and the surroundings we can, however, make Taylor expansions and derive simple expressions for ∇, i.e. for the temperature gradient, which we need to determine.

6.5.2   *The temperature gradient for large energy exchange*

Large energy exchange means $\Gamma \ll 1$ and inefficient convection. In this case one obtains by means of Taylor expansions that

$$\nabla = \nabla_{\text{rad}} - \frac{9}{4}\Gamma^2(\nabla_{\text{rad}} - \nabla_{\text{ad}}) \qquad (6.40)$$

and

$$\bar{v} = \frac{R_g T}{8\mu} Q\Gamma_0(\nabla_{\text{rad}} - \nabla_{\text{ad}}) \qquad (6.41)$$

All quantities on the right-hand side of equation (6.40) can be calculated locally if T and P_g are known. The temperature stratification as a function of the gas pressure can therefore be calculated by a straightforward integration of equation (6.40). We see that the gradient ∇ approaches the radiative gradient ∇_{rad} for small values of Γ, i.e. for large energy exchange and very inefficient convective energy transport.

6.5.3   *The temperature stratification for small energy exchange*

Small energy exchange means $\Gamma \gg 1$. By a Taylor expansion in terms of $1/\Gamma$ we find

$$\nabla = \nabla_{\text{ad}} + \frac{4}{9\Gamma}(\nabla_{\text{rad}} - \nabla_{\text{ad}}) \qquad (6.42)$$

and

$$\bar{v}^3 = \frac{R_g T}{18\mu} Q \frac{1}{\Gamma_0} (\nabla_{rad} - \nabla_{ad}) \tag{6.43}$$

Again the temperature stratification can be obtained by a straight-forward integration of equation (6.42). For very large values of Γ, i.e. for very little energy exchange and **very efficient convective energy transport, the temperature stratification becomes nearly adiabatic**.

Of course the actual temperature gradient can never be exactly adiabatic or flatter than the adiabatic one, even for the most active convection. Were this to happen the stellar layer would become stable against convection. The motions would stop, the convective energy transport would decay and the temperature gradient would increase until convection could set in again. For very efficient convection the temperature gradient can approach the adiabatic one and be nearly indistinguishable from it, but it can never actually become adiabatic or flatter than that.

For intermediate values of Γ one has to follow the iterative procedure by Böhm-Vitense (1953) or the method described by Kippenhahn, Weigert and Hofmeister (1967).

6.6   Temperature stratification with convection in stellar interiors

In stellar interiors the densities are very high and we therefore expect very efficient convective energy transport in convectively unstable regions. For the pressure scale height we also find large values because of high temperatures. A value $l \approx H = R/10$ appears to be a reasonable estimate.

According to equation (6.16) the amount of convective energy transport can be described by

$$\pi F_c = \rho c_p T \sqrt{g} \left(\frac{l}{2}\right)^2 \left(\frac{\Delta \operatorname{grad} T}{T}\right)^{3/2} \tag{6.44}$$

with

$$\Delta \operatorname{grad} T = \frac{dT}{dz} - \left(\frac{dT}{dz}\right)_{bu}$$

In deep layers the gas bubbles are large and κ_{cm} is very large because of the high density. The gas bubbles are therefore optically very thick, the energy exchange is minute ($\Gamma \gg 1$) and $(dT/dz)_{bu} = (dT/dz)_{ad}$.

Solving equation (6.44) for Δ grad T gives us the difference between the adiabatic gradient and the average temperature gradient. We find

$$(\Delta \text{ grad } T)^{3/2} = \pi F_c T^{3/2}/[c_p \rho T \sqrt{g}(l/2)^2] \qquad (6.45)$$

We know that the convective flux must be less than the total energy flux, i.e. $F_c < F$. We can therefore say

$$(\Delta \text{ grad } T)^{3/2} < \pi F T^{3/2}/[c_p \rho T \sqrt{g}(l/2)^2] \qquad (6.46)$$

which gives an upper limit for Δ grad T.

The total flux to be transported can be calculated from the observed surface flux and the radius. It must be

$$\pi F = \sigma T_{\text{eff}}^4 \frac{R^2}{r^2} \approx 4\sigma T_{\text{eff}}^4 \quad \text{if} \quad r \approx R/2 \qquad (6.47)$$

We use $c_p = 5R_g/2\mu$ and $\mu = 0.6$, and find

$$c_p \rho T = \frac{5}{2} \frac{R_g T}{\mu} \rho \approx \frac{5}{2} P_g$$

and

$$g = \frac{GM_r}{r^2} \approx \frac{4g(\text{surface})}{2} \approx 2g(\text{surface}) \approx 10^4 \qquad (6.48)$$

if $M_r \approx \frac{1}{2}M$.

For some interior point we very roughly estimate $P_g \sim 10^{15}$ dyn cm^{-2}, $T \approx 10^7$ K and

$$\left(\frac{l}{2}\right)^2 \approx \frac{R^2}{4 \times 10^2} \approx 10^{18} \text{ cm}^2$$

We then derive for solar values of T_{eff}, R and g

$$(\Delta \text{ grad } T)^{3/2} \leqslant \frac{2 \times 10^{11} \times T^{3/2}}{5 \times 10^{15} \times 10^2 \times 10^{18}} \leqslant 10^{-24} \times T^{3/2}$$

and $(\Delta \text{ grad } T)^{3/2} \leqslant 10^{-14}$ or Δ grad $T \leqslant 10^{-10}$. This has to be compared with the average temperature gradient in the star which can be estimated to be $\overline{dT/dz} \approx T_c/R \approx 10^7/10^{11} \approx 10^{-4}$. Δ grad T is much smaller than dT/dr. We therefore find in stellar interiors

$$\Delta \text{ grad } T = \frac{dT}{dz} - \left(\frac{dT}{dz}\right)_{\text{ad}} < 10^{-6} \times \frac{dT}{dz} \quad \text{or} \quad \frac{dT}{dz} \sim \left(\frac{dT}{dz}\right)_{\text{ad}} \qquad (6.49)$$

As discussed previously the temperature gradient can never be exactly equal to the adiabatic one, because then convection would stop, but within our calculation accuracy the actual temperature gradient and the adiabatic one are equal.

In the top layers of the outer convection zone the radiative energy exchange is rather large (the optical depths of the rising and falling gas bubbles are not very large) and the temperature stratification deviates by only a little from the radiative equilibrium stratification. The deeper we go into the outer convection zone, the larger the density and the characteristic length and the more efficient is the convective energy transport. The deviations from the radiative equilibrium stratification becomes larger and finally the adiabatic temperature gradient is approached. **The temperature increase with depth is reduced by the convective energy transport.**

6.7   Convective overshoot

In the preceding sections we saw that a rising bubble will be accelerated by the buoyancy force as long as it has a higher temperature than its surroundings. Generally we find that its temperature difference with respect to the surroundings keeps increasing as long as the radiative temperature gradient inwards is greater than the adiabatic one, i.e. as long as the layer in the star is convectively unstable, i.e. up to the top of the convection zone. It therefore reaches the top of the convectively unstable zone with a positive ΔT and an upward velocity. Because of the positive ΔT it is still accelerated when it reaches the boundary of the unstable region and thus overshoots (see Fig. 6.6). Beyond this boundary its ΔT will, however, decrease, the acceleration becomes smaller, and finally zero when ΔT has decreased to zero. Only when ΔT becomes negative will the bubble experience a negative acceleration and finally come to rest. We may therefore always expect a fair amount of overshoot. The exact amount depends on how fast the ΔT decreases, which depends on how stable the overlying stable region is, and also on the efficiency of the energy exchange between the bubble and the surroundings, which becomes important at the upper boundary of the outer convection zone.

The overshoot of the convective bubbles into the overlying stable region is the reason why we see granulation in the convectively stable solar photosphere.

The corresponding situation for the lower boundary of the convection zone is shown schematically in Fig. 6.7.

6.8   Convective versus radiative energy transport

Without detailed calculations we can estimate the largest amount of convective energy transport possible. As we just saw, for extremely efficient convection the actual temperature gradient ∇ approaches the adiabatic one which can be small but never 0. For a given temperature gradient the radiative energy flux is given by equation (3.7) and is $F_r \propto dT/dz > (dT/dz)_{ad}$. In the case of radiative equilibrium the temperature gradient is given by $(dT/dz)_{rad}$ and $\pi F_r = L/4\pi r^2 = \pi F$. We therefore find that

$$\frac{F_r}{F} > \frac{\nabla_{ad}}{\nabla_{rad}} \tag{6.50}$$

If $\nabla_{ad} \ll \nabla_{rad}$ then $F_r \ll F$, but even then we never transport all the energy by convection. In large fractions of the convection zone the radiative temperature gradient is only somewhat larger than the adiabatic one; we therefore usually still transport a large fraction of the energy by

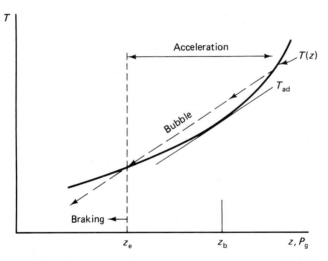

Fig. 6.6. Shows schematically the temperature stratification $T(z)$ of the stellar atmosphere at the surface of the convection zone (thick solid line) and the adiabatic temperature stratification (thin solid line). The upper boundary of the convection zone occurs at the depth z_b where $dT/dz = (dT/dz)_{ad}$. A rising bubble may follow the path indicated by the dashed line. At the upper boundary of the convection zone at z_b the bubble still has $\Delta T > 0$ and is still being accelerated up to the depth z_e, for which $\Delta T = 0$. Due to its inertia it still continues to move upwards but obtains now $\Delta T < 0$ and is braked. It finally comes to rest at a point $z_0 < z_e$, i.e. much higher than the upper boundary of the convection zone. Energy exchange reduces the overshoot distance.

radiation. Only in the top layers of the outer hydrogen convection zone where $\nabla_{rad} \gg \nabla_{ad}$ do we find that almost all of the energy is transported by convection. For these layers we can then give a close upper limit for the convective velocities by writing $\pi F_c = \sigma T_{eff}^4$ and making use of equations (6.16) and (6.17) (see also Volume 2).

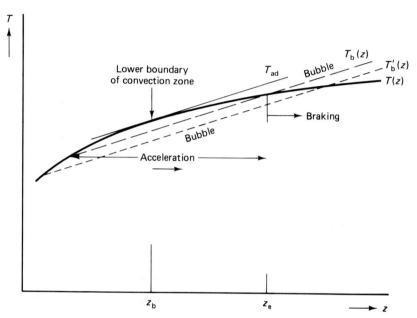

Fig. 6.7. Shows schematically the temperature stratification $T(z)$ of the atmosphere at the bottom of the convection zone (thick solid line). Also shown is the adiabatic temperature stratification T_{ad} (thin solid line). The lower boundary of the convection zone occurs at the depth z_b where $dT/dz = (dT/dz)_{ad}$.

A falling bubble follows an adiabatic temperature change $T_b(T)$ (long dashes). It arrives at the lower boundary of the convection zone z_b with a negative ΔT and continues to be accelerated down to the depth z_e where its $\Delta T = 0$. Due to its inertia it continues to fall but now obtains a positive ΔT in the convectively stable region. Its motion is braked and it finally comes to rest. A bubble falling from a larger height follows the line $T'_b(t)$ (dotted line) and overshoots even further.

7

Depths of the outer convection zones

7.1   General discussion

Our next question is: how deep are the outer convection zones and which stars have them? The answer is not simple because the depth of the unstable region depends on the efficiency of the convective energy transport. Qualitatively we can understand this in the following way.

In Fig. 7.1 we show schematically the radiative equilibrium temperature stratification in the outer layers of a convectively unstable zone. As we pointed out earlier convection sets in when hydrogen begins to ionize. For such temperatures ∇_{ad} decreases and κ increases. The adiabatic temperature gradient returns to its normal value of $\nabla_{ad} \approx 0.4$ when hydrogen and helium are completely ionized. We are then dealing again with a monatomic gas in which any heat energy supplied is used to increase the kinetic energy, i.e. the temperature, of the gas. We find that for complete ionization the absorption coefficient also decreases, because the electrons which could absorb in the continuum are removed, at least from hydrogen and helium. For the remaining atoms we saw that for high temperatures the average κ generally decreases for deeper layers as $\kappa \propto T^{-2.5}$. In a qualitative way we can therefore say that the lower boundary of the outer convection zones occur at the layer where the ionization of the most abundant elements is nearly complete.

The degree of ionization can be calculated from the Saha equation:

$$\frac{N^+}{N} = \frac{2u^+}{u} \frac{(2\pi m_e kT)^{3/2}}{h^3} \exp\left(-\chi_{ion}/kT\right) \frac{1}{n_e} \tag{7.1}$$

where N^+ indicates the number of ionized particles and N the number of neutral particles. (Of course, we can write down a similar equation for N^{++}/N^+.) u^+ and u are the so-called partition functions for the ion and the atom respectively, which describe the distribution of the electrons over the different energy levels of the ion or atom (see Volume 2; m_e = electron mass).

The larger T is, the greater is the ratio N^+/N which means the greater the degree of ionization. The larger the electron density n_e, the smaller the ratio N^+/N. For larger electron densities we need higher temperatures to reach the same degree of ionization achievable for lower n_e with a lower temperature.

From the Saha equation we can calculate the relation between temperature and pressure for which the ionization of helium is about, say, 99 per cent complete. For increasing pressures we need higher temperatures to achieve this degree of ionization. In Fig. 7.1 we have schematically indicated the relation between pressure and temperature for which the ionization of helium is 99 per cent complete.

We have also indicated the radiative equilibrium temperature stratification which would be obtained if convective energy transport were unimportant in spite of convective instability. At the point where this curve intersects with the complete ionization line the temperature reaches

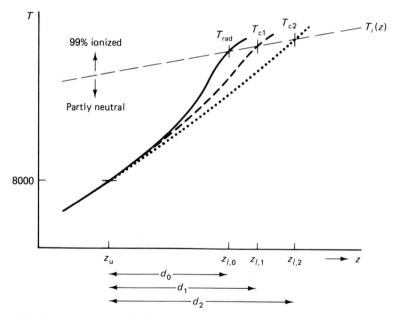

Fig. 7.1. The radiative equilibrium temperature stratification T_{rad} plotted schematically as a function of depth z (solid line). Also shown are qualitatively the temperature stratifications for inefficient convective energy transport, T_{c1} (dashed line), and for efficient convection, T_{c2} (dotted line). Hydrogen and helium are nearly completely ionized for temperatures above a line $T_i(z)$ in the $T(z)$ plane as indicated by the long dashes. (For deeper values with higher P_g, larger T are required according to the Saha equation; see Volume 2.) The intersection of the temperature stratification with this line shows the lower boundary of the hydrogen convection zone. For more efficient convection the boundary occurs at a deeper layer.

values for which convective instability seizes. The intersection point at depth $z_{l,0}$ therefore marks the lower boundary of the convectively unstable region. For radiative equilibrium stratification we find that this happens at $P_g = P_{g0}$ and $T = T_0$. The depth of the convectively unstable region is given by d_0 if the stratification remains the same as in radiative equilibrium.

We now take convective energy transport into account, which decreases the necessary radiative energy transport and therefore leads to a smaller temperature increase with increasing depth. The temperature stratification is then as indicated by the dashed line in Fig. 7.1. The line for complete ionization intersects the temperature stratification line at a higher pressure because the temperature required for ionization is now only reached at a greater depth $z_{l,1}$. For this higher pressure we also need a somewhat higher temperature than in the case of radiative equilibrium stratification. The depth of the convection zone is therefore given by $d_1 > d_0$. For very efficient convective energy transport the required radiative flux is even smaller and the temperature increase is still smaller. We may then find a temperature stratification as shown by the dotted line. The intersection of the ionization line with the temperature stratification line occurs at even larger depth, $z_{l,2}$, and the depth of the convection zone increases even further to d_2. The largest possible depth occurs for adiabatic temperature stratification. **Thus convection extends the region of convective instability.**

The differences in depth obtained for different efficiencies of convective energy transport are not small. When Unsöld (1931) studied the depth of the solar outer convection zone using radiative equilibrium stratification he found a depth of 2000 km, while Biermann (1937), assuming adiabatic temperature stratification, determined a depth of 200 000 km.

The calculated convective energy transport increases with increasing values for the assumed characteristic length l; therefore the calculated depth of the convection zone also increases with increasing values for l. We could in principle determine the proper value for l if we could measure the depth of the convection zone. We will come back to this question in Section 7.4.

7.2   Dependence of convection zone depths on T_{eff}

In Chapter 3 we derived $F_r = \frac{4}{3}(dB/d\tau)$ with $\tau = \int_0^z \kappa_{cm} \, dz$. For stellar atmospheres with $F_r = F = \sigma T_{\text{eff}}^4/\pi = \text{const.}$, we find by integration $B = \frac{3}{4}F\tau + \text{const.}$ With $B = \sigma T^4/\pi$ we derive

$$T^4 = \tfrac{3}{4}T_{\text{eff}}^4 \tau + T^4(\tau = 0), \qquad (7.2)$$

with $T^4(\tau = 0) \sim \tfrac{1}{2}T_{\text{eff}}^4$ (see Volume 2).

For $T_{\text{eff}} \sim 5800$ K, as in the sun, hydrogen starts to ionize for $T \sim 6000$ K and $P_g \sim 10^5$ dyn cm^{-2}. For higher gas pressures higher temperatures are needed. For stars with lower T_{eff} the temperature increases more slowly with optical depth τ as seen from equation (7.2). For the lower temperatures the absorption coefficients κ_{cm} are smaller; therefore a given optical depth τ and a given temperature is reached only at greater depth z where the gas pressures are larger. **For lower effective temperatures we therefore find the onset of hydrogen ionization at deeper layers with higher pressures.** For higher pressures the convective energy transport becomes more efficient. In Fig. 7.2 we plot schematically the temperature versus pressure stratifications for main sequence stars of different T_{eff}. For lower

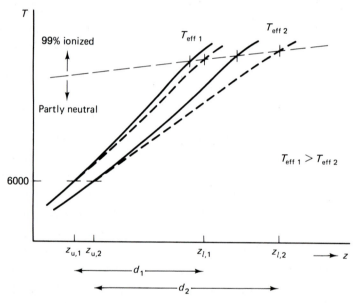

Fig. 7.2. Shows schematically the temperature stratifications of two stars with different T_{eff}, T_{eff}(star 1) $> T_{\text{eff}}$(star 2). The solid lines show the radiative equilibrium temperature stratifications. For the lower T_{eff} the temperature increases more slowly with depth z. Also shown are the temperature stratifications with convective energy transport (short dashes). We also have plotted schematically the line above which hydrogen and helium are completely ionized (long dashes). Convective instability sets in for $T \sim 6000$ K. For the hotter star this occurs at $z_{u,1}$; for the cooler star it occurs at a larger depth $z_{u,2}$. For star 1, with the larger T_{eff}, convection sets in at a lower pressure and lower density than for star 2. Convective energy transport is therefore less efficient in star 1 than in star 2. The lower boundary of the convective zone of star 1 occurs at a higher layer with $z = z_{l,1}$; that for star 2 occurs at $z_{l,2}$. Star 1 has a convection zone with depth d_1, while for star 2 the depth is $d_2 > d_1$.

effective temperatures we find higher pressures in the convection zones which leads to more efficient convection and smaller temperature gradients. The ionization is completed only at layers with higher temperatures and higher pressures at larger depths. The depths of the outer convection zones increase. For very cool main sequence stars the outer convection zones may perhaps extend down to the center. The question is still under debate.

For stars with higher effective temperatures we consequently find lower pressures in the atmosphere because a given optical depth occurs at a higher layer due to the higher atmospheric absorption coefficients. At a given optical depth we also have higher temperatures because of the higher T_{eff} (see equation 7.2). The ionization of hydrogen therefore starts higher in the atmosphere. For early F stars the convectively unstable region may start as high as at $\tau = 0.2$. For such low pressures the convective energy transport becomes inefficient. For $T_{eff} \sim 7600 \, \text{K}$ at spectral type A9 or F0 the temperature stratification no longer differs from the radiative equilibrium stratification and the convection zones become very thin.

7.3 Dependence of convection zone depths on chemical abundances

If element abundances change, the depth of the convection zone also changes. For higher helium abundances the pressure in the atmosphere increases as does the electron pressure. The ionization of helium will therefore be completed only at a higher temperature, deeper in the atmosphere. We expect a deeper convection zone for helium-rich stars.

For cool metal-poor stars the electron pressure in the atmospheres is decreased and the absorption coefficient is therefore decreased. We find at a given optical depth a larger gas pressure. The pressure in the outer convection zones is also larger and convective energy transport therefore becomes more efficient. For thin convection zones the depth increases. For thick convection zones it appears, however, that another effect becomes important. In Chapter 4 we discussed how for temperatures above a few hundred thousand degrees the absorption of heavy elements is most important. Metal-rich stars therefore have a larger κ. This increase in κ increases the radiative temperature gradient and thereby increases the convective instability. For metal-poor stars this absorption is reduced and hence the convective instability in this region may be reduced. Spite and Spite (1982) found that for metal-poor stars the depths of the convection zones are smaller than for solar abundance stars.

7.4   The lithium problem

As we discussed in the previous sections the depth we derive for the outer convection zones depends on the assumed efficiency of convective energy transport, i.e. on the value assumed for the characteristic length l. If we find a way to determine the depth of the convection zone we can determine the correct value for l. This is of course important to know because in many areas of the HR diagram the computed stellar structure depends very sensitively on the assumed value of l. Fortunately nature has provided us with some information about the depth of this unstable zone. Because of the convective motions the material in the convection zone proper is thoroughly mixed in a rather short time; for the sun it turns out to be of the order of one month. Since the rising material has a positive ΔT when it arrives at the upper boundary of the convection zone it will overshoot beyond the upper boundary of the convection zone as discussed in Section 6.7. At the bottom of the convection zone the same arguments apply to the falling gas which will overshoot into the stable region below the convection zone proper (see Fig. 6.7). The overshoot velocities are expected to decrease probably exponentially with increasing depth, possibly with a scale of the order of a pressure scale height or less. Accurate calculations are not available at this moment. In any case we may expect that there is some slow mixing of the material in the convection zone with the material below the convection zone proper. This is important when we look at the observed abundances of lithium in stellar atmospheres. Let us look at the example of the sun. For temperatures above 2×10^6 K the Li^6 nuclei are destroyed by nuclear reactions with protons, i.e.

$$Li^6 + H^1 \rightarrow He^4 + He^3$$

Li^7 burns at somewhat higher T, namely $T > 2.4 \times 10^6$ K by the reaction

$$Li^7 + H^1 \rightarrow 2He^4$$

Clearly the lithium we see at the solar surface is not in surroundings at 2 million degrees, but if, due to convection, the whole outer regions down to the layers with temperatures of 2 million degrees are mixed, then Li^6 is mixed down to layers which are hot enough to destroy it. This means that if the convection zone reaches down to such high temperature regions Li^6 is destroyed in a time which is very short in comparison with the solar age of 4.5×10^9 years. Since we do not observe any Li^6 in the solar spectrum, the convection zone must extend to layers with temperatures of 2 million degrees. On the other hand we do observe a weak Li^7 line, but the solar Li^7 abundance is reduced by about a factor of 100 compared with the abundance observed in meteorites and in the most Li-rich stars, the very young T Tauri stars. It appears that the convection zone proper does not quite reach down to the layer with a temperature of 2.4 million degrees though some very slow, probably overshoot, mixing with surface material still occurs, which reduces the Li^7 abundance on a time scale of 10^9 years. Other mixing processes may also have to be considered.

Assuming $l = H$ one calculates that the convection zone proper reaches down to 2 million degrees; for $l = 2H$ it would extend to 3 million degrees, which appears to be incompatible with the presence of any Li^7 in the solar atmosphere. For $l \approx H$ we find the right conditions: Li^6 is destroyed on a short time scale, but only slow overshoot mixing reaches the layer where Li^7 is destroyed.

In other stars the lithium abundances can also be studied. No Li^6 has been detected in any of the main sequence stars, but Li^7 abundances have been determined for many stars. Ages of stars can be determined if they occur in clusters (see Section 8.10). For other main sequence stars ages can be determined from the slight increase in luminosity (see Sections 14.2 and 13.1). These studies show that for all main sequence stars cooler than F0 stars the Li is destroyed on a long time scale, which seems to decrease somewhat for cooler stars because of the increase in the depth of the convection zone. However, we are in trouble if we want to understand the observations for F stars, for which we calculate that convection zones do not reach down to layers with temperatures of 2 million degrees, nor do we expect slow overshoot mixing down to layers with 2.4 million degrees. For these stars convective mixing cannot be the explanation for lithium destruction. The observation by Boesgaard and Tripicco (1986) that in some young cluster stars the surface lithium depletion is largest for stars with spectral types around F5 presents us with another puzzle. Perhaps rotation induced mixing or diffusion (see Volume 1) is important.

8

Energy generation in stars

8.1   Available energy sources

So far we have only derived that, because of the observed thermal equilibrium, the energy transport through the star must be independent of depth as long as there is no energy generation. The energy ultimately has to be generated somewhere in the star in order to keep up with the energy loss at the surface and to prevent the star from further contraction. The energy source ultimately determines the radius of the star.

Making use of the condition of hydrostatic equilibrium we estimated the internal temperature but we do not yet know what keeps the temperature at this level. In this chapter we will describe our present knowledge about the energy generation which prevents the star from shrinking further.

First we will see which energy sources are possible candidates. In Chapter 2 we talked about the gravitational energy which is released when the stars contract. We saw that the stars must lose half of the energy liberated by contraction before they can continue to contract. We might therefore suspect that this could be the energy source for the stars.

The first question we have to ask is how much energy is actually needed to keep the stars shining. Each second the sun loses an amount of energy which is given by its luminosity, $L = 3.96 \times 10^{33}$ erg s^{-1}, as we discussed in Chapter 1. From the radioactive decay of uranium in meteorites we can find that the age of these meteorites is about 4.5×10^9 years. We also find signs that the solar wind has been present for about the same time. These observations together provide convincing evidence that the sun has existed for at least that long and that it has been shining for that long. Studies of different layers of sedimentation in old rock layers give us information about the climate on Earth over several billions of years; these indicate that the climate on Earth has not been vastly different from the present climate. From this we gather that the sun must have been shining with about the same energy for a major fraction of the time that the Earth has existed, namely for most of 4.5×10^9 years. We may then conclude that the Sun over its lifetime has lost an amount of energy which is roughly given by

$$E_\odot = L \times 4.5 \times 10^9 \text{ years} = 4 \times 10^{33} \text{ erg s}^{-1} \times 4.5 \times 10^9 \times 3 \times 10^7 \text{ s}$$
$$= 5.4 \times 10^{50} \text{ erg}$$

In Chapter 2 we calculated that the *gravitational energy release* due to the contraction of the Sun to its present size is

$$E_g \approx \frac{GM^2}{R} \approx \frac{7 \times 10^{-8} \times 4 \times 10^{66}}{7 \times 10^{10}} = 4 \times 10^{48} \text{ erg}$$

The available energy from this energy source is a factor of 100 too small to keep the sun shining for 4.5×10^9 years. From the gravitational energy the sun could survive for only 10^7 years (the Kelvin–Helmholtz time scale, as we said earlier). Gravitational energy probably supplied the solar luminosity for the very young sun but it cannot now be the solar energy source.

Let us make a rough estimate whether any *chemical energy source* could supply enough energy. For the most optimistic estimate let us assume that the sun would consist of hydrogen and oxygen in the best possible proportions such that all solar material could be burned to water vapor. A water molecule has a molecular weight of $18m_H = 18 \times 1.66 \times 10^{-24}$ g $= 3.5 \times 10^{-23}$ g. In the burning process $(2 \times 10^{33} \text{ g})/(3.5 \times 10^{-23} \text{ g}) = 6 \times 10^{55}$ molecules could then be formed. Suppose that in each formation of a molecule an energy of the order of 10 eV $= 10^{-11}$ erg is liberated. The total energy available from such a chemical process (which fuels the best rockets) would then be 6×10^{44} erg. This amount of energy is still a factor of 10^4 smaller than the available gravitational energy, and would last only for about 5000 years. We have to look for much more powerful energy sources. The most powerful energy source we can think of is nuclear energy.

8.2   Nuclear energy sources

Looking at the masses of atoms we find that an energy of about 1 per cent of the rest mass of the protons can be gained by combining hydrogen into heavier nuclei. This can easily be seen from Table 8.1 in which we have listed the atomic weights of some of the more important elements. The atomic weight of hydrogen is 1.008 172, defining the atomic weight of O^{16} as 16. In these mass units the atomic weight of the He^4 isotope is 4.003 875, while four times the atomic weight of hydrogen is

Table 8.1. *Isotopic masses and mass defects*

A	Isotope	$\mu(C^{12})$	$\mu(O^{16})$	$A \times \mu_H$	Δm	$\Delta m/m$
1	H^1	1.007 852	1.008 172			
4	He^4	4.002 603	4.003 875	4.032 688	0.0288	0.0071
12	C^{12}	12.000 000	12.003 815	12.098 064	0.0942	0.0078
16	O^{16}	15.994 915	16.000 000	16.130 752	0.1307	0.0081
20	Ne^{20}	19.992 440	19.998 796	20.163 440	0.1646	0.0082
56	Fe^{56}	55.934 934	55.952 717	56.457 632	0.5049	0.0089

4.032 688. **In the process of making one He^4 nucleus out of four protons, 0.0288 mass units are lost.** This mass difference has been transformed into energy according to Einstein's relation

$$E = \Delta mc^2 \tag{8.1}$$

The mass fraction which is transformed into energy in this process is then 0.0288/4 = 0.007, or 0.7 per cent. We know that three-quarters of the stellar mass is hydrogen. If stars can find a way to combine all these hydrogen atoms to helium a fraction of 0.5 per cent of the stellar mass could be transformed into energy. If the stars could manage to combine all their hydrogen and helium into Fe^{56} nuclei then an even larger fraction of the mass could be transformed into energy. From Table 8.1 we calculate that the mass of 56 protons is 56.457 63 mass units while Fe^{56} has a mass of 55.9527. In this fusion process a mass fraction of 0.9 per cent of the stellar mass for a pure hydrogen star could be transformed into energy, which means 20 per cent more than by the fusion of hydrogen into helium. This energy could be available to heat the material in the stellar nuclear reaction zones from where the energy can then be transported to the surface where it is lost by radiation.

Let us estimate whether this energy source is sufficient to provide the energy needed to keep the sun shining for 5×10^9 years. The available energy would then be

$$\Delta mc^2 = 0.009 \times Mc^2 \approx 0.01 \times (2 \times 10^{33} \text{ g}) \times (9 \times 10^{20} \text{ cm}^2 \text{ s}^{-2})$$
$$= 2 \times 10^{52} \text{ erg}$$

We calculated previously that the sun has radiated about 6×10^{50} erg during its lifetime up until now. From the available nuclear energy sources it could then in principle live 30 times as long as it has lived already if it could convert all its mass into Fe by nuclear processes. Actually, the stars leave the main sequence already when they have consumed only about 10 per cent of their fuel. The sun will stay on the main sequence only for about 10^{10} years. No more efficient energy source than nuclear energy can be thought of unless the sun shrank to a much smaller radius. We can estimate how small the radius of the sun would have to become if the gravitational energy release were to exceed the nuclear energy which is in principle available. We then have to require that

$$E_{\text{grav}} = \frac{GM^2}{R} > 2 \times 10^{52} \text{ erg}$$

or

$$\frac{7 \times 10^{-8} \times 4 \times 10^{66}}{R} > 2 \times 10^{52}$$

or

$$R < \frac{7 \times 10^{-8} \times 4 \times 10^{66}}{2 \times 10^{52}} \approx 1 \times 10^7 \text{ cm} = 100 \text{ km}$$

This means it would have to shrink to a density

$$\rho = \frac{M}{\frac{4}{3}\pi R^3} \geqslant \frac{2 \times 10^{33}}{4 \times 10^{21}} = 0.5 \times 10^{12} \text{ g cm}^{-3} = 5 \times 10^5 \text{ tonne cm}^{-3}$$

Such densities can be obtained, but only in neutron stars. As a normal star the sun has no larger energy source than its nuclear energy. The question now is whether nuclear reactions can actually take place in stellar interiors.

8.3 The tunnel effect

In order for two nuclei to fuse they have to get very close together. Each nucleus, i, has, however, a positive nuclear charge $Z_i e$. There are therefore strong repulsive forces between two nuclei. In order to bring two charged nuclei together to a small distance r against the Coulomb force they need an energy

$$E_{\text{Coulomb}} = \frac{Z_1 Z_2 e^2}{r} \tag{8.2}$$

where $Z_1 e$ and $Z_2 e$ are the nuclear charges of the two particles involved.

For fusion the particles have to get sufficiently close to allow nuclear forces to act, which means they have to get closer than about 10^{-13} cm. With $r = 10^{-13}$ cm we calculate

$$E_{\text{Coulomb}} = \frac{Z_1 Z_2 e^2}{10^{-13}} \approx 2 \times 10^{-6} \text{ erg} \approx 1000 \text{ keV}$$

In a star the kinetic energies of the particles is the only energy available to overcome these Coulomb barriers. The average kinetic energy of a particle is

$$E_{\text{kin}} = \tfrac{3}{2}kT = \tfrac{3}{2} \times 1.4 \times 10^{-16} \times 10^7 \text{ erg} = 2 \times 10^{-9} \text{ erg}$$

for a temperature of 10^7 K. Obviously the particles are short by a factor 1000 of the necessary kinetic energy. We know from the Maxwell velocity distribution that there are always particles which have larger than average energies E but their number decreases as $N(E) \propto e^{-E/kT}$. For an energy which is roughly a factor of 1000 larger than the average energy the number $N(E)$ therefore decreases by a factor of roughly e^{-1000} or 10^{-430}. How many particles are then in the solar interior with such an energy? The total number of particles N in the Sun is approximately $N = \text{mass}/m_{\text{H}} = 2 \times 10^{33}/1.7 \times 10^{-24} \approx 10^{57}$. The chance of finding even *one* particle with a kinetic energy large enough to overcome the Coulomb barrier is essentially zero. How then can the charged particles ever get close enough for fusion? Quantum mechanics has shown that there is a very small but nonzero possibility that even a particle with rather low kinetic energy can penetrate the Coulomb barrier because of the tunnel effect, as we will see below.

In Fig. 8.1 we have plotted qualitatively the Coulomb potential $V(r)$ in the neighborhood of a charged nucleus; i.e. $V(r)$ is the energy which a charged particle has to have in order to get to the distance r near the nucleus. Very close to the nucleus (i.e. $r < r_2 \sim 10^{-13}$ cm) nuclear forces start to attract the other nucleus. Very close to the nucleus the particle therefore gains energy when coming closer, and the potential decreases steeply. The particle falls into the so-called potential well. In order to calculate quantum mechanically the probability of finding the particle at a distance r we have to solve the Schrödinger equation for the ψ function for the oncoming particle in the potential of the nucleus.

The Schrödinger equation for ψ is

$$\frac{\partial^2 \psi}{\partial r^2} + \frac{8\pi^2}{h^2}(E_{\text{kin}} - V)\psi = 0 \tag{8.3}$$

The solutions are exponentials. As long as $E_{\text{kin}} > V(r)$ the exponent is imaginary and the solution for the ψ function is a wave function with amplitude A_0 (see Fig. 8.1). In the region $r < r_1$ where $V(r) > E_{\text{kin}}$ the exponent becomes negative, the character of the wave function changes to an exponentially decreasing function with the exponent being proportional to $-\sqrt{(V - E_{kin})}$. For $r < r_2$ we find again $E_{kin} > V(r)$ (see Fig. 8.1) and the ψ function. Its derivative must match the value of the exponential at $r = r_2$. For $r < r_2$ the ψ function has an amplitude A_i which is much smaller than A_0. The amplitude for the ψ function for $r < r_2$ determines the probability of finding the particle close to the nucleus, which means it determines the penetration probability $P_p(E)$ for the particle to penetrate the Coulomb barrier, or to 'tunnel' through the Coulomb barrier. For small values of E_{kin} the distance $r_1 - r_2$ becomes

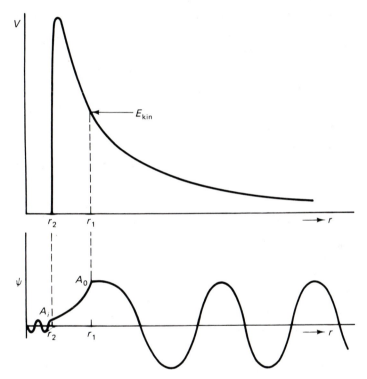

Fig. 8.1. Shows schematically the Coulomb potential $V(r)$ (solid line) for one nucleus in the neighborhood of another nucleus at $r = 0$. $V(r)$ is the energy which is needed by a particle coming from infinity to get to the distance r_1 working against the repulsive Coulomb force. For distances $r < r_2$ nuclear forces attract the approaching particle and the potential becomes negative. This is the so-called potential well.

    If the particle has a kinetic energy E_{kin} then according to classical theory it could not approach closer than r_1. For $r > r_1$ where $E_{kin} > V$ the ψ function for the particle is a sine wave with amplitude A_0 according to quantum theory. For $r < r_1$ the ψ function is an exponential function decreasing towards smaller r. At r_2 it has a value A_i. For $r < r_2$ we find again $E_{kin} > V$ and the ψ function is again a wave but now with amplitude A_i. For larger E_{kin} the distance $r_1 - r_2$ decreases and the ratio A_i/A_0 increases.

large and therefore $A_i = \psi(r_2)$ becomes very small; the penetration probability is then extremely small. For larger energies the penetration probability is larger. We calculate

$$P_{\mathrm{p}}(v) \propto \exp\left(\frac{-4\pi^2 Z_1 Z_2 e^2}{h}\frac{1}{v}\right) \tag{8.4}$$

The Coulomb barrier increases with increasing $Z_1 Z_2 e^2$, and therefore the tunneling probability decreases exponentially with this term. Because of this exponential, nuclear reactions between nuclei with low values of Z_i are the only ones which can occur at relatively low temperatures. This means that reactions between protons seem to have the best chance.

We also see the factor $1/v$ in the exponent telling us that the penetration probability becomes larger for larger velocities when the exponent becomes smaller. As we saw above for larger velocities, i.e. for larger kinetic energies of the particles, the width of the potential wall becomes narrower, the decay of the ψ function less steep and the amplitude of the ψ function in the center of the Coulomb barrier remains larger, and so the penetration probability is larger.

The chance of penetrating the Coulomb barrier alone does not yet completely describe the probability of a nuclear reaction. Nuclei may come close together but not react if there are other problems for the reaction. For instance, a nuclear reaction between two protons, which gives the smallest value for the exponent in equation (8.4), has an extremely small chance of taking place because there is no stable helium nucleus with two positive nuclear charges and no neutrons in it. The only possible nucleus with mass number 2 is the deuteron, which has one proton and one neutron. For this nucleus to form one of the two protons has to become a neutron by emitting a positron and this has to happen while the two protons are close together in the potential well. The decay of a proton into a neutron and a positron is a process due to the weak interaction and is therefore a rare event. The probability of a deuteron forming during a collision between two protons is therefore quite small; nevertheless it does happen even for relatively low temperatures, like 4 million degrees. However, the reaction cross-section is so small that it can never be observed in the laboratory. For $T = 1.4 \times 10^7$ K and $\rho X \approx 100$ (X = fraction or hydrogen by mass), as we have in the center of the sun, it would take 1.4×10^{10} years $= 5 \times 10^{17}$ s before a given proton, which moves around, reacts with any other proton. (This is the time after which all protons in the center have reacted with another proton, after which time the hydrogen in the center is exhausted. This is then the lifetime of the sun on the main sequence.)

8.4 The proton–proton chain

It turns out that the easiest nuclear reaction is the reaction between a deuteron and a proton. Both nuclear charges are 1, and the deuteron and proton can directly form a stable helium nucleus with mass 3. This reaction can happen with temperatures around 2 million degrees. For the conditions in the center of the Sun it takes only 6 seconds for a deuteron to react with a proton in the hydrogen gas. There are, however, only a few deuteron nuclei available, so the number of these processes is very small; they are nevertheless very important once a proton–proton reaction has taken place and formed a deuteron.

The probability for proton–proton fusion itself does not depend on the velocity of the particles. Since the product $Z_1 Z_2$ is quite small the probability for the tunnel effect is not very strongly dependent on temperature for temperatures around 10^7 K. In order to calculate the number of proton–proton reactions per cm^3 s, we have to sum the contributions from protons with different velocities, taking into account the number of particles, $N(v_x)$, within any given velocity interval dv_x as given by the Maxwell velocity distribution

$$dN(v_x) = N \frac{1}{\pi} \exp\left[\frac{-mv_x^2}{2kT}\right] dv_x \tag{8.5}$$

Here x is the direction along the line connecting the two particles. N is the total number of particles per cm^3. The final probability for any nuclear reaction of the kind i, the reaction rate R_i, is then given by the product of the Maxwell velocity distribution with the penetration probability for a given velocity and the nuclear reaction probability which is often expressed as a reaction cross-section C_i. We thus have

$$R_i = \int_{v_x=0}^{v_x=\infty} P_p(Z_1, Z_2, v_x) C_i \, dN(v_x) \tag{8.6}$$

In Fig. 8.2 we show qualitatively the two probability functions $dN(v_x)$ and $P_p(v_x)$. The function $dN(v_x)$ decreases exponentially with increasing v_x, while the penetration probability increases exponentially with v_x^{-1}. The product of these two functions has a fairly sharp peak at the velocity v_G, the so-called Gamow peak, which determines the value of the integral in equation (8.5). Most reactions take place between particles with relative velocity v_G. For many purposes we can therefore perform calculations as if all the nuclear reactions of a given kind occur with the same velocity, v_G, the velocity at which we find the Gamow peak.

Once two protons have combined to form a deuteron, D^2, further reactions quickly follow. At 'low' temperature the following reactions happen, called the PPI chain:

$$\text{PPI: } H^1 + H^1 \rightarrow D^2 + e^+ + \nu \qquad (1.4 \times 10^{10} \text{ years})$$
$$D^2 + H^1 \rightarrow He^3 + \gamma \qquad (6\,\text{s})$$

and then

$$He^3 + He^3 \rightarrow He^4 + H^1 + H^1 \qquad (10^6 \text{ years}) \qquad (8.7)$$

where the reaction times given apply to a single nucleus traversing the solar interior with $T \sim 1.4 \times 10^6$ K and $\rho \approx 100$ g cm^{-3} and a solar chemical composition. The ν indicates the emission of a neutrino.

The reaction chain can also end in the following ways:

$$\text{PPII: } He^3 + He^4 \rightarrow Be^7 + \gamma$$
$$Be^7 + e^- \rightarrow Li^7 + \nu$$
$$Li^7 + H^1 \rightarrow He^4 + He^4 \qquad (8.7)$$

or

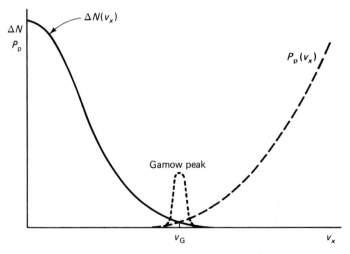

Fig. 8.2. Shows schematically the velocity distribution for the velocity component v_x, i.e. $\Delta N(v_x)$ (solid line). Also shown schematically is the velocity dependence of the penetration probability $P_p(v_x)$ (long dashes). The product of the two functions has a sharp peak, the so-called Gamow peak (dotted line, not drawn to scale).

$$\text{PPIII:} \quad \text{Be}^7 + \text{H}^1 \rightarrow \text{B}^8 + \gamma$$
$$\text{B}^8 \rightarrow \text{Be}^8 + e^+ + \nu$$
$$\text{Be}^8 \rightarrow 2\text{He}^4 \tag{8.8}$$

The relative importance of the PPI and PPII chains depends on the relative importance of the reactions of He^3 with He^3 in PPI as compared to the reactions He^3 with He^4 in PPII. For $T > 1.4 \times 10^7$ K, He^3 prefers to react with He^4. For lower T the PPI chain is more important.

The PPIII ending is never very important for energy generation, but it generates high energy neutrinos. For the temperatures in the solar interior the PPIII chain is very temperature sensitive because it involves reactions of nuclei with $Z_i = 4$. The number of high energy neutrinos generated by these reactions is therefore very temperature dependent. If we could measure the number of these high energy neutrinos they would give us a very sensitive thermometer for the central temperatures of the sun. We will come back to this problem in Section 18.6, when we discuss the solar neutrino problem.

Since the neutrinos can escape freely from the sun without interacting with the solar material their energy is lost for the solar heating. As the neutrinos generated in the three chains have different energies different fractions of the total energy are lost in this way for different interior temperatures. For the three chains the different neutrino energy losses amount to 1.9 per cent for the PPI chain, 3.9 per cent for the PPII chain and 27.3 per cent for the PPIII chain.

8.5   The carbon–nitrogen cycle

For higher temperatures reactions involving nuclei with somewhat higher Z_i can become important if the reaction cross-sections are especially large, as in resonance reactions. Resonance reactions occur when the kinetic energy of the captured nucleon coincides with an excited energy level of the combined nucleus; this makes the formation of the combined nucleus very easy.

In the carbon cycle the following reactions constitute the main cycle:

$$\text{C}^{12} + \text{H}^1 \rightarrow \text{N}^{13} + \gamma \qquad (10^6 \text{ years})$$
$$\text{N}^{13} \rightarrow \text{C}^{13} + e^+ + \nu \qquad (14 \text{ min})$$
$$\text{C}^{13} + \text{H}^1 \rightarrow \text{N}^{14} + \gamma \qquad (3 \times 10^5 \text{ years})$$

$$N^{14} + H^1 \rightarrow O^{15} + \gamma \qquad (3 \times 10^8 \text{ years})$$
$$O^{15} \rightarrow N^{15} + e^+ + \nu \qquad (82\,\text{s})$$
$$N^{15} + H^1 \rightarrow C^{12} + He^4 \qquad (10^4 \text{ years}) \qquad (8.9)$$

The time estimates refer to the conditions given above, corresponding to the solar interior ($T = 1.4 \times 10^7$, $\rho \sim 100$). The net effect of the whole chain of reactions is the formation of one He^4 from four H^1, while C^{12} is recovered at the end. The total number of C, N and O nuclei does not change in the process.

As in the PP cycle different endings are possible. N^{15} can also react in the following way:

$$N^{15} + H^1 \rightarrow O^{16} + \gamma$$
$$O^{16} + H^1 \rightarrow F^{17} + \gamma$$
$$F^{17} \rightarrow O^{17} + e^+ + \nu$$
$$O^{17} + H^1 \rightarrow N^{14} + He^4 \qquad (8.10)$$

N^{14} then enters the main cycle at line 4. This gives some increase in the production of He^4 but this ending is only 4×10^{-4} times as frequent as the main cycle. The main importance of this bi-cycle is a change in the O^{16} and O^{17} abundances (see Section 8.8 and Chapter 13).

8.6   The triple-alpha reaction

For still higher temperatures, namely $T \sim 10^8$ K, a reaction

$$Be^8 + He^4 \rightarrow C^{12} + 2\gamma + 7.6\,\text{MeV}$$

can take place. For the CNO cycle we had $Z_1 Z_2 \leqslant 7$, now $Z_1 Z_2 = 8$. Why do we need much higher temperatures for this reaction? The problem is that Be^8 is unstable to fission. It is very short lived. So how do we get Be^8 for this reaction? In the fission process energy is gained, 95 keV. This means in order to make Be^8 from two He^4 nuclei energy has to be put in, exactly 95 keV. This is the inverse of an ionization process where energy is gained by recombination and energy has to be put in for the ionization. Yet there are always a few ions around; the number of ions is increasing with increasing temperature, according to the Saha equation. Similarly, we find an increasing number of Be^8 nuclei (in this case the combined particle) with increasing temperature.

For a temperature $T = 10^8$ K the average kinetic energy $\frac{3}{2}kT$ is around 2×10^{-8} erg. Since 1 erg $= 6.24 \times 10^{11}$ eV we find $\frac{3}{2}kT \approx 12$ keV. The energy needed for the formation of Be^8 is only about a factor of 8 higher.

Let us look at the situation for the ionization of hydrogen which for atmospheric pressures takes place when $T \approx 10\,000$ K. At this temperature the average kinetic energy $\frac{3}{2}kT$ is about 2×10^{-12} erg, or just about 1 eV. The ionization energy for hydrogen is 13.6 eV, yet the hydrogen is ionized. Therefore with a recombination energy of 95 keV and a mean kinetic energy of about 12 keV, we can still expect to find a sufficient number of Be^8 in the equilibrium situation

$$\text{He}^4 + \text{He}^4 \Leftrightarrow \text{Be}^8 - 95 \text{ keV} \tag{8.11}$$

The number of Be^8 nuclei is always quite small but still large enough such that some reactions

$$\text{Be}^8 + \text{He}^4 \rightarrow \text{C}^{12} + 2\gamma + 7.4 \text{ MeV} \tag{8.12}$$

can take place. The net effect of (8.11) and (8.12) is obviously to make one C^{12} from three He^4 nuclei, i.e. from three alpha particles. This reaction is therefore called the triple-alpha reaction. The net energy gain is 7.4 MeV $-$ 95 keV $\sim$ 7.3 MeV.

8.7   Element production in stars

Observations tell us that low abundances of heavy elements are seen in very old stars. We actually observe stars whose heavy element abundances are reduced by a factor up to 10^4 as compared to the sun. On the other hand, young stars have all heavy element abundances which are not very different from solar abundances. This suggests that the abundances of heavy elements increased during the formation and evolution of the galaxy and that probably the first generation of stars had no heavy elements at all. This is also suggested by cosmological studies which show that during the high temperature phase of the big bang there was not enough time to make any elements more massive that He^4 except for very few Li nuclei. The *very* few Be^8 nuclei which probably were present were far too few and the time much too short to make C^{12} in the triple-alpha reaction.

The first generation of stars probably consisted only of hydrogen and helium. The CNO cycle could therefore not occur, only the PP chains. It was an important discovery of Hoyle that the triple-alpha reaction could work in evolved stars because these stars have about 10^5 years available to make C^{12} in the triple-alpha reactions. Without this reaction we would have no heavy elements, which means we would not exist. Once C^{12} is formed then further captures of helium nuclei or alpha particles are possible for increasing temperatures in the interiors of massive stars; then nuclei up to Fe^{56} can form. Clearly not all observed elements and isotopes can be made by adding alpha particles. We also need captures of protons and neutrons by the heavy nuclei, in order to build all the observed elements up to uranium with a mass of 238 atomic mass units. In the course of the late stages of evolution of massive stars conditions can be identified under which the different processes can take place, though there are still some important unsolved problems with respect to the details of the processes.

Basically we feel certain that the elements more massive than He^4 were all made in the hot interiors of evolved stars. Without the triple-alpha reaction there would be only hydrogen and helium gas; no solid planets, no life.

8.8   Comparison of different energy generation mechanisms

From Table 8.1, we know that the total amount of energy to be gained from the triple-alpha reaction can only be around 10 per cent of what can be gained from the fusion of four H^1 into one He^4. In each fusion process $4H^1 \rightarrow He^4$ we gain ~ 26.2 MeV from the PP cycle and 25.0 MeV from the CNO cycle (the difference being due to the different amounts of neutrino energy lost). Considering that it takes three times as much mass to form one C^{12} as it takes to form one He^4 we find again that the energy gain from the reaction $3He^4 \rightarrow C^{12}$ is 7.3 MeV/3 − 2.4 MeV per He^4 nucleus as compared to 25 MeV from the CNO cycle, or just about 10 per cent. A star with a given mass and luminosity can therefore live on this energy source at most 10 per cent of the time it can live on hydrogen burning.

As the stars on the main sequence live on hydrogen fusion, they can stay there much longer than at any other place in the color magnitude diagram. That is the reason why we find most stars on the main sequence. On the other hand we also see that the hot and massive O and B stars can remain on the main sequence only for a much shorter time than the cooler stars because they use up their available fuel much faster; the cooler stars are much more careful in using their fuel and can therefore survive much longer.

Fig. 8.3 shows how much energy is generated by the different mechanisms for different temperatures. Interior temperatures for the sun and for some typical stars are also indicated.

For M dwarfs, i.e. cool stars, the energy generation is mainly due to the PP chain. For the sun both the PP and the CNO cycles contribute, but the PP chain is more important. For Sirius A, an A0 V star with $T_{\text{eff}} \sim 10\,000$ K the CNO cycle is more important than the PP chain. Triple-alpha reactions, also called helium burning, are not important for 'normal' stars on the main sequence except for the most massive ones. Very high interior temperatures are needed.

We saw earlier that the temperature dependence of nuclear reactions is actually an exponential one, but since the PP chain is only important for low temperatures and the CNO cycle mainly for high temperatures, i.e. $T \geqslant 1.7 \times 10^7$ K, it is often convenient to approximate the actual temperature dependence by a simple power law. In this approximation we find that the energy generation per gram second by the PP chain, ε_{pp}, is given by

$$\varepsilon_{\text{pp}} = \varepsilon_{\text{p}} \rho X^2 \left(\frac{T}{10^6} \right)^{\nu} \tag{8.13}$$

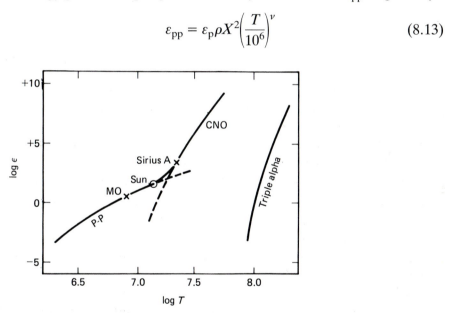

Fig. 8.3. The energy generation per gram is shown as a function of temperature for the different nuclear processes. Solar interior densities and abundances are used for the proton–proton (PP) and CNO reactions. For the triple-alpha reactions densities higher by a factor 10^3 were used because these reactions only occur for $T > 10^8$ K when much higher densities are found in stellar interiors. Also indicated are the conditions in the centers of some main sequence stars. For low temperatures the proton–proton chain is the most efficient mechanism even though it generates only small amounts of energy. For the sun the PP chain is still more important than the CNO cycle. (Adapted from Schwarzschild 1958b.)

with $\nu \sim 4$. X is the abundance of hydrogen in mass fraction. For energy generation by the CNO cycle, ε_{cc}, we find

$$\varepsilon_{cc} = \varepsilon_c \rho X Z(C, N)\left(\frac{T}{10^6}\right)^\nu \tag{8.14}$$

where ν depends somewhat on the temperature range which we are considering but is usually $\nu \sim 16$ when the CNO cycle is important. The difference in ν for the PP and the CNO reactions expresses the fact that the CNO cycle is much more temperature sensitive than the PP chain.

$Z(C, N)$ gives the abundance of C or N in mass fractions. The number of fusion processes per cm^3 depends on the number of collisions between H^1 and H^1 in the PP chain, or between C^{12} and H^1 or N^{14} and H^1 in the CNO cycle. ε_{pp} per cm^3 therefore depends on $n(H)n(H) \propto \rho^2 X^2$, and the ε_{cc} per cm^3 depends on $n(H)n(N) \propto (\rho X)\rho Z(N) \propto \rho^2 XZ(N)$ or $\varepsilon_{cc} \propto n(H)n(C) \propto \rho^2 XZ(C)$. ε_p and ε_c in equations (8.13) and (8.14) are constants and are determined by the reaction rates. The energy production per gram equals the energy production per cm^3 divided by the number of grams per cm^3, i.e. divided by ρ, which leads to (8.13) and (8.14).

8.9   Equilibrium abundances

In equations (8.6) and (8.9) we have given approximate values for the times that are needed on average for one reaction to occur if we shoot one heavy particle into a hydrogen gas at a density of $\rho X = 100$ and $T = 14 \times 10^6$ K.

Suppose the first reaction takes 10^6 years to form an N^{13} nucleus from any given C^{12} atom and one H^1. The N^{13} will decay in 14 minutes. This means that we have very little chance of finding an N^{13} nucleus. It will become C^{13} immediately and the next N^{13} will be formed only 10^6 years later. C^{13} remains there for 3×10^5 years, so we have a much better chance of finding it than we have of finding N^{13}, but still, after C^{13} is gone, we will have to wait 7×10^5 years before the next one is formed. No matter how fast the reaction C^{13} + H^1 $\rightarrow$ N^{14} + γ goes we still have to wait more than 10^6 years for the first N^{14} to be created because it takes so long for the first reaction. Once this N^{14} is formed it will on average stay around for a long time, namely 3×10^8 years, before it is transformed into O^{15}. In the meantime 300 additional N^{14} have been formed by the first reactions. The abundance of N^{14} is increased. **The relative abundances of the particles**

involved in the CNO cycle are changed in the process, depending on the reaction rates. The abundance of N^{14} increases at the expense of the C^{12} abundance. The less probable a given reaction is, the larger will be the abundance of the element which is waiting to make this reaction. On the other hand we also see that the more N^{14} nuclei there are, the larger is the chance that *one* of them will react and become O^{15}. If the relative abundance of N^{14} to C^{12} has increased to 300 (C^{12} decreases, N^{14} increases) then each time one new N^{14} is created by the first reactions in the CNO cycle one N^{14} also disappears. The abundance of N^{14} no longer changes. After about 3×10^8 years N^{14} has reached its equilibrium abundance.

The equilibrium values are reached if the number of destruction processes per unit of time for a given nucleus equals the number of creation processes per unit of time. This means for instance for N^{14} abundances in equilibrium that

$$\frac{dN^{14}}{dt} = C^{13} \times \text{reaction rate}(C^{13} + H^1) - N^{14} \times \text{reaction rate}(N^{14} + H^1)$$

$$= 0$$

With the reaction rate being proportional to (reaction time)$^{-1}$ we have according to the numbers given in equations (8.9)

$$\frac{C^{13}}{3 \times 10^5} - \frac{N^{14}}{3 \times 10^8} = 0 \quad \text{or} \quad \frac{N^{14}}{C^{13}} = \frac{3 \times 18^8}{3 \times 10^5} = 1000 \text{ for equilibrium}$$

Or for the ratio of C^{12} to C^{13}

$$\frac{C^{12}}{10^6} = \frac{C^{13}}{3 \times 10^5} \quad \text{and} \quad \frac{C^{12}}{C^{13}} = \frac{10^6}{3 \times 10^5} = 3.3$$

which leads to

$$\frac{N^{14}}{C^{12}} = \frac{N^{14}}{C^{13}} \frac{C^{13}}{C^{12}} = 300$$

As the reaction rates depend on the temperature the equilibrium abundance ratios also depend somewhat on the temperature. In any case we expect an increase in the nitrogen abundance and a decrease in the carbon abundance in those stellar regions where the CNO cycle has been operating even if it contributes little to the energy generation. We therefore expect an increase in the C^{13}/C^{12} ratio also in the solar interior. In the solar atmosphere this ratio is only about 0.01 but it can reach the value of 0.3 in those regions where the CNO cycle has been operating long enough at temperatures around 1.4×10^7 K. For the equilibrium between the C^{12} to C^{13} abundances to be established we need about 10^6 years for $T \sim 1.4 \times 10^7$ K.

If we find unusually high N^{14} and C^{13} abundances together we suspect that we see material which once has been in those regions of a star in which the CNO cycle has been operating. We do indeed find such anomalous abundances in the atmospheres of most red giants and supergiants. We will come back to this point in Chapter 18.

8.10   Age determination for star clusters

We saw in Section 8.1 that nuclear energy is the only energy source in stars which is large enough to supply the stars with enough energy to replenish what they lose by radiation. The available amount of nuclear energy can be estimated from the mass defect, that is the amount of mass lost when combining four hydrogen nuclei, or protons, to one helium nucleus, or one alpha particle. We saw that 0.7 per cent of the mass is lost in this process and converted to energy. When we combine this with the knowledge (to be shown later) that the stars change in appearance after they have transformed about 10 per cent of their hydrogen mass into helium we can easily estimate how long the sun and all the main sequence stars can remain in their present configuration. We find that the total nuclear energy available on the main sequence is given by

$$E_{\text{nuclear}} = \Delta mc^2 = 0.1M \times 0.007c^2 = 6.3 \times 10^{17}M \text{ erg} \qquad (8.15)$$

This overestimates the available energy by a small amount because a minor percentage is lost by means of escaping neutrinos, and also because some of the mass is helium to start with.

In order to determine the time t for which this energy supply lasts we have to divide it by the amount of energy which is used per unit of time; this is given by the luminosity of the star. We find

$$t = \frac{E_{\text{nuclear}}}{L} = \frac{6.3 \times 10^{17}M \text{ erg}}{L \text{ [erg s}^{-1})} \qquad (8.16)$$

For the sun we have $M_\odot = 2 \times 10^{33}$ g, $L_\odot = 4 \times 10^{33}$ erg s^{-1}. With this we obtain

$$t_\odot = \frac{6.3 \times 10^{17} \times 2 \times 10^{33}}{4 \times 10^{33}} \text{ s} = 3.15 \times 10^{17} \text{ s} \sim 10^{10} \text{ years}$$

The total lifetime of the sun on the main sequence is about 10^{10} years. The sun has already spent half of its lifetime on the main sequence.

For other stars we have to insert mass and luminosity into equation (8.17). It is more informative if we use directly the observed mass–luminosity relation

$$\frac{L}{L_\odot} \propto \left(\frac{M}{M_\odot}\right)^{3.5} \tag{8.17}$$

We then derive

$$\frac{t}{t_\odot} \propto \frac{M}{M_\odot}\frac{L_\odot}{L} \propto \frac{M}{M_\odot}\left(\frac{M_\odot}{M}\right)^{3.5} = \left(\frac{M}{M_\odot}\right)^{-2.5} \tag{8.18}$$

For a star whose mass is ten times the solar mass, a B1 V star, the lifetime on the main sequence is shorter by a factor of about 300 than that of the sun. The B1 star evolves away from the main sequence after about 3×10^7 years or 30 million years. For the most massive stars the lifetime on the main sequence is only about 1 million years! The most massive stars which we see now must have been formed within the last million years!

What does this mean with respect to the appearance of cluster color magnitude diagrams?

In Fig. 8.4 we show schematic color magnitude diagrams as expected for different ages, looking only at main sequence stars. For a very young cluster, $10^4 <$ age $< 10^6$ years, we expect to see the very massive stars on the main sequence. Their contraction time is about 10^4 years. In the course of their contraction half of the gravitational energy released has remained in the star as thermal energy. At some point their internal energy and temperature has become high enough such that nuclear fusion processes

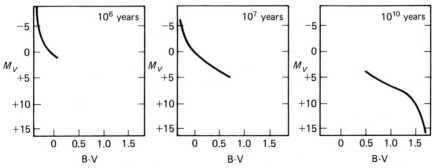

Fig. 8.4. Expected main sequences in the color magnitude diagrams for clusters of different ages are shown schematically. The age is indicated for each curve.

can start. The energy loss is then made up by nuclear energy generation. The star stops contracting. Its radius no longer changes. In the color magnitude diagram it remains in the same position as long as the nuclear energy generation can resupply the surface energy lost by radiation. The star essentially does not change its appearance (except that it slowly increases its brightness by about 0.5 magnitude). Since the stars remain at this stage for the largest part of their lifetimes, this is the configuration in which we find most of the stars. This is the main sequence stage of evolution for the stars. The most massive stars, the so-called O stars, can remain at this stage only for 10^6 years as we estimated. After 10^6 years they disappear from the main sequence and become supergiants, as we shall see in Chapter 15. Of course, after 10^6 years stars with small masses like the sun have not even reached the main sequence. After 10^6 years the main sequence only extends down to the A stars with $B - V \sim 0.10$, but the hottest, O stars, have already started to disappear. After 10^7 years the O stars have all disappeared from the main sequence, which extends now at the upper end to B0 stars and at the lower end to stars with $B - V \sim 0.6$, i.e. to G stars like the sun. After 10^8 years the main sequence extends at

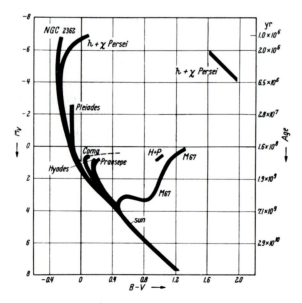

Fig. 8.5. Shows a superposition of color magnitude diagrams for different galactic clusters and the globular cluster M67. The distance moduli have been determined in such a way that the main sequences match at the lower end. For the Pleiades the top of the main sequence ends for early B stars indicating an age of about 10^8 years, while for M67 the main sequence stops for early G stars ($B - V \sim 0.60$) like the sun. Its age must be about 10^{10} years. H + P stands for Hyades and Praesepe. Adapted from Arp (1958).

the lower end to the K stars with $B - V \sim 0.9$ but at the top it ends at stars with $\sim 8\,M_\odot$, the early B stars. For 10^9 years it stops at the top at $B - V \sim 0$, at the early A stars, and at 10^{10} years it extends only to early G stars like the sun.

The upper termination point of the main sequence therefore gives us an age estimate for such clusters. In Fig. 8.5 we show the superposition of the color magnitude diagrams of a number of galactic clusters. We can easily derive that the Pleiades star cluster must have been formed about 100 million years ago and the globular cluster M67 must be roughly 10^{10} years old. This is, of course, only a rough estimate since we have used only the approximate mass–luminosity relation, equation (8.17).

9

Basic stellar structure equations

9.1   The temperature gradient

We now have available all the basic physics and also the basic equations to discuss the temperature and pressure stratification in stars from which the theoretical relations between mass and chemical composition as input parameters, on the one hand, and the radius, luminosity and effective temperatures on the other hand, can be derived.

As discussed before the observed fact that most of the stars do not change their temperatures measurably in time tells us that they must be in *thermal equilibrium*. This requires that the luminosity $L(r)$, which is the total amount of energy passing through each spherical shell at the distance r from the center, has to be constant throughout the star, except in those layers where energy is generated. In such layers the newly generated energy also has to be transported out, else the energy generating layer would have to heat up. Thermal equilibrium therefore requires that

$$\frac{\mathrm{d}L(r)}{\mathrm{d}r} = 4\pi r^2 \rho(r)\varepsilon(r) \tag{9.1}$$

$L(r)$ includes radiative, convective and possibly conductive energy transport. $\varepsilon(r)$ includes all kinds of energy generation. If the star is contracting or expanding we have to include gravitational energy release (or consumption) in $\varepsilon(r)$ (see Chapter 20).

In Chapter 8 we found for nuclear energy generation

$$\varepsilon = \varepsilon_0 \rho T^\nu \tag{9.2}$$

where $\nu \approx 4$ for the proton–proton chain and where $\nu \approx 16$ for the CNO cycle.

Equations (9.1) and (9.2) determine the luminosity L as a function of r.

The temperature gradient is determined by the radiative flux F_r according to

$$\frac{4}{3}\frac{dB}{d\bar{\tau}} = F_r \qquad (9.3)$$

where $F_r = L_r/4\pi r^2$ and L_r is the radiative luminosity, i.e. the radiative energy flux through each spherical shell.

In the case of radiative equilibrium $L_r = L$ and we find for the *temperature stratification in radiative equilibrium*

$$\frac{dT}{dr} = -\frac{L(r)}{4\pi r^2}\frac{\kappa_{cm}}{T^3}\frac{3}{16\sigma} \qquad (9.4)$$

The dependence of the Rosseland mean opacity κ_{cm} on T and P_g and on element abundances was discussed in Chapter 4. We use here for the average κ per gram of material

$$\kappa_g = \kappa_0\rho^\alpha T^{-\beta} \qquad (9.5)$$

where $\kappa_0 \propto Z$ for the bound-free and bound-bound transitions.

For Kramers' opacity law we have $\alpha = 1$ and $\beta = 3.5$. The numerical calculations favor $\alpha = 0.5$ and $\beta = 2.5$, as we saw from Fig. 4.3.

We derived that in *the convection zone* we can usually expect the adiabatic temperature and pressure relation (except in the upper parts of the outer hydrogen and helium convection zones); this leads to the relation (see equation (5.35))

$$\frac{dT}{dr} = \nabla_{ad}\frac{T}{P}\frac{dP}{dr} \qquad (9.6)$$

for *deep convection zones*. For the outer, low density layers we have to use the equations discussed in Chapter 7.

9.2   The pressure gradient

The observed fact that the stars do not change their radii also tells us that the stars must be in *hydrostatic equilibrium*, as expressed by the hydrostatic equation (see equation (2.12))

$$\frac{dP}{dr} = -\rho\frac{GM_r}{r^2} \quad \text{or} \quad \frac{dP_g}{dr} = -\frac{GM_r}{r^2}\rho \qquad (9.7)$$

if only the gas pressure is important. In general we have to use $P = P_g + P_r +$, etc., where P_r is the radiation pressure. Other pressures may be due to turbulence or magnetic fields.

The thermal equilibrium (9.1), also called the energy equation, and the hydrostatic equilibrium equations (9.7) are actually all we have in order to determine the stellar structure. The additional equations which we have to use only help to evaluate the quantities occurring in these equilibrium equations.

In order to determine M_r as a function of the radius r, we have to describe how the mass inside radius r increases with increasing r. When increasing r by the amount dr the mass M_r increases by the amount dM_r inside the spherical shell with radius r and thickness dr, which has a volume $4\pi r^2$ dr and a density ρ. This means

$$\frac{dM_r}{dr} = 4\pi r^2 \rho \qquad (9.8)$$

Equations (9.1), (9.4) (or 9.6), (9.7) and (9.8) provide four equations for the five unknowns $P_g(r)$, $T(r)$, $\rho(r)$ and $M_r(r)$. We need an additional equation, relating P_g, ρ and T, namely the equation of state. Once we discuss stellar interiors of very high density we have to take degeneracy into account (see Chapter 16). For the following discussions we can, however, use the well-known relation for an ideal gas

$$P_g = \frac{\rho}{\mu} R_g T \qquad (9.9)$$

If the radiation pressure P_r is important we have

$$P_r = \frac{1}{3} \frac{4\sigma}{c} T^4 \quad \text{and} \quad P = P_g + P_r \qquad (9.10)$$

Equation (9.9) serves to eliminate ρ, which leaves us with the four unknowns T, P_g, M_r and L.

9.3   The boundary conditions

The unique solution of the four first-order differential equations (9.1), (9.4), (9.7) and (9.8) requires the knowledge of four boundary conditions. We can easily write down two boundary conditions at the surface, namely at $\bar{\tau} = \frac{2}{3}$ we must have a temperature $T = T_{\text{eff}}$ (see Volume 2). Since generally the temperature and pressure decrease very fast at the surface this condition can usually be simplified to $T(R) = 0$. For giants and supergiants, for which the pressure decreases only slowly, we have to be more careful, however. We also know that at the surface the pressure has to go to zero, i.e. $P_g(R) = 0$.

For the center of the star where $r = 0$ we can also easily formulate two boundary conditions; namely for $r = 0$ the mass inside r has to be zero, which means $M_r(0) = 0$. We also know that there is no energy generated inside $r = 0$, and we therefore have $L(0) = 0$. We thus have two boundary conditions at the outside and two at the inside

$$T(R) = 0, \quad P_g(R) = 0, \quad M_r(0) = 0 \quad \text{and} \quad L(0) = 0 \qquad (9.11)$$

where R is determined by the condition that $M_r(R) = M$.

The fact that we have only two boundary conditions on each side causes problems for the solution. If we start the integration of the differential equations from the outside we have to assume L and R and hope that by the time we reach the point $r = 0$ we do indeed find $L = 0$ and $M_r = 0$. We could try to find the correct solution by trial and error. It turns out, however, that this is impossible because the solutions all diverge at the center. Why? Because the r occurs in the denominator of equations (9.3) and (9.7). Minute errors in M_r and L_r cause the pressure and temperature gradients to be infinite for $r \to 0$. We could try to start the integration from the inside where we have to guess $P(0)$ and $T(0)$. We could then try again to find the correct solution by trial and error, but we encounter the same problem: the solutions diverge at the surface, where the temperature and pressure go to zero. We never match the surface boundary conditions. One possibility is to start from both ends, determining the starting conditions in such a way that the solutions from both ends match smoothly at some intermediate point. 'Match smoothly' must then mean that T, P, L and M_r and their derivatives are continuous at this point. This method was extensively used by several authors, especially by M. Schwarzschild who developed clever forms of matching conditions and succeeded in computing evolutionary tracks in this way even before large computers were available. Modern calculations are based on the method developed by L. Henyey. It introduces all four boundary conditions simultaneously. We will outline the Henyey method in Chapter 12.

9.4 Dimensionless structure equations

Schwarzschild discovered that the amount of computational work could be greatly reduced by introducing new dimensionless variables. Using these variables we will be able to gain some basic insights into the dependence of stellar structure on mass, chemical composition and age, necessarily with some simplifying assumption, mainly that we are dealing with so-called homologous stars and with stars of homogeneous chemical composition.

The new dimensionless variables are chosen to be

$$x = \frac{r}{R}, \quad q = \frac{M_r}{M}, \quad f = \frac{L(r)}{L}, \quad t = \frac{T}{T_0} \quad \text{and} \quad p = \frac{P_g}{P_0} \quad (9.12)$$

The constants T_0 and P_0 are chosen in such a way as to simplify the basic structure equations. This is the case if

$$T_0 = \frac{\mu GM}{RR_g} \quad \text{and} \quad P_0 = \frac{GM^2}{4\pi R^4}$$

where μ is the mean atomic weight of the homogeneous star. For a given star T_0 and P_0 are constants.

With these transformations the basic differential equations obtain the following forms:

$$\frac{dp}{dx} = -\frac{p}{t}\frac{q}{x^2} \quad (9.13)$$

to replace (9.7). For equation (9.8) determining M_r we find now

$$\frac{dq}{dx} = \frac{p}{t}x^2 \quad (9.14)$$

For the temperature gradient in radiative equilibrium we obtain

$$\frac{dt}{dx} = -C\frac{f}{x^2}\frac{p^{\alpha+1}}{t^{\alpha+\beta+4}} \quad (9.15)$$

if $\kappa_g = \kappa_0 \rho^\alpha T^{-\beta}$.

For adiabatic stratification with $c_p/c_v = \frac{5}{3}$ we derive

$$p = Et^{2.5} \quad (9.16)$$

The energy equation now becomes

$$\frac{df}{dx} = Dp^2 t^{\nu-2}x^2 \quad (9.17)$$

Here

$$C = C_0 \frac{\kappa_0}{\mu^{\beta+4}}\frac{LR^{\beta-3\alpha}}{M^{\beta+3-\alpha}} \quad \text{with} \quad C_0 = \frac{3}{16\sigma}\left(\frac{R_g}{G}\right)^{\beta+4}\left(\frac{1}{4\pi}\right)^{\alpha+2} \quad (9.18)$$

We also have

$$D = D_0 \varepsilon_0 \mu^\nu \frac{M^{\nu+2}}{LR^{\nu+3}} \quad \text{with} \quad D_0 = \left(\frac{G}{R_g}\right)^\nu \frac{1}{4\pi} \tag{9.19}$$

if $\varepsilon = \varepsilon_0 \rho T^\nu$. The constant E is derived to be

$$E = 4\pi K G^{3/2} (\mu/R_g)^{5/2} M^{1/2} R^{3/2} \tag{9.20}$$

where the constant K determines the adiabat which is followed by the temperature stratification, namely

$$T^{5/2} = P_g/K \tag{9.21}$$

C_0 and D_0 are elementary constants, **while C, D and E are constant for each star**, C and D depending only on M, L and R. This is very important. For these new variables the boundary conditions are also well defined: at the surface, for $x = 1$, we must have $q = 1$, $f = 1$, $t = 0$ and $p = 0$. At the center for $x = 0$, we must require that $q = 0$ and $f = 0$. For our four unknown functions we have now six boundary conditions to satisfy. We have four well-defined boundary conditions at the surface and can therefore start integration from the surface and find a well-defined solution for given values of C and D. We do, however, have two necessary boundary conditions in the center which also have to be satisfied. For arbitrary values of C and D we cannot expect this to happen. We can, however, try to choose C and D in such a way that the interior boundary conditions are also satisfied. Since we have two boundary conditions in the center these conditions determine the possible values of C and D uniquely. **There is only *one* pair of C and D with which the inner boundary conditions can be satisfied.** With this one pair of values for C and D, which are well-defined numbers, we find *a unique solution* for $q(x)$, $t(x)$, $p(x)$ and $f(x)$ for the four basic differential equations (9.13)–(9.15) and (9.17). **This means we have the same solution for all stars in radiative equilibrium. C and D must also be the same for all stars in radiative equilibrium** with $P \approx P_g$ and with ε and κ given by equations (9.2) and (9.5) respectively.

Because of the numerical problems discussed earlier, this solution also cannot be obtained by integration from outside in. Integrations also have to be started from both sides and the constants D and C are determined by the condition of a smooth transition from the inside to the outside solution.

If we are dealing with convective stars the constant E replaces the constant C. **All completely convective stars also have one unique solution if $P \sim P_g$ and (9.2) and (9.5) hold.**

Such stars then all show the same dependence of q, t, p, f on the radius variable x. They are also *homologous stars*. Unfortunately real stars are not all homologous stars because they have convection zones of various depths as we saw in the previous chapters and for hot stars radiation pressure and electron scattering becomes very important. Nevertheless homologous stars are reasonable approximations to stars of intermediate mass in which the convection zones are not very extended, like for instance the A, F, G and possibly B stars.

10

Homologous stars in radiative equilibrium

10.1 The dependence of stellar parameters on mass

In this section we will see what we can learn about homologous stars. As we found in the previous chapter all homologous stars in radiative equilibrium with $P \sim P_g$ and for which equations (9.2) and (9.5) hold have the same solutions for $t(x)$, $p(x)$, $q(x)$, $f(x)$ with the *same values for the constants C and D* occurring in the equations (9.15) and (9.17). These constants contain functions of M, R and L. They must therefore tell us something about these quantities and their relation for the different stars.

We realize that by multiplying the equations for C and D we derive an equation which no longer contains the luminosity L; it gives a relation between R and M only:

$$CD = \text{const.} \frac{R^{\beta - 3\alpha} M^{\nu+2}}{R^{\nu+3} M^{\beta - \alpha + 3}} \tag{10.1}$$

or

$$R^{\nu + 3 - \beta + 3\alpha} \propto M^{\nu - 1 - \beta + \alpha} \tag{10.2}$$

If the carbon cycle is the main energy source, which means if $\nu = 16$, and Kramers' opacity law holds we find

$$R \propto M^{12.5/18.5} \quad \text{or roughly} \quad R \propto M^{2/3} \tag{10.3}$$

For $\beta = 2.5$ and $\alpha = 0.5$ we obtain

$$R \propto M^{13/18.0} = M^{0.72} \tag{10.3a}$$

With $T \propto M/R$ (see Chapter 3), we find $T \propto M^{1/3}$. **With increasing mass the interior temperature of the stars must increase** which leads to the transition from the **proton–proton chain as the energy source for low mass stars** to the **CNO cycle for high mass stars.** We cannot expect, however, the effective temperature to obey this mass dependence because we have used an opacity law which does not hold for the stellar photospheres, and also equation (9.2) is not true for optically thin atmospheric layers. But, of course, T_{eff} also increases with M.

With this relation between R and M we find now from the equation for C that for Kramers' opacity law

$$L \propto \frac{M^{5.5}}{R^{0.5}} \propto \frac{M^{5.5}}{M^{1/3}} \approx M^{5.2} \tag{10.4}$$

For $\beta = 2.5$ and $\alpha = 0.5$ we have

$$L \propto M^{4.3} \tag{10.4a}$$

This is the mass–luminosity relation for idealized homologous stars in radiative equilibrium for which the CNO cycle is the main energy source. A and F stars with their small inner and outer convection zones best match the conditions. We want to emphasize that the mass–luminosity relation follows mainly from C, which means mainly from the equation of energy *transport*. The radius, which depends on the energy generation, enters only with a very small power.

We can now estimate the dependence of T_{eff} on the mass using the mass dependences of the luminosities and radii. We ·have $T_{eff}^4 \propto L/R^2 \propto M^{5.2}/M^{1.36} \approx M^{3.8}$, for Kramers' opacity which gives $T_{eff} \propto M^{0.96}$. This increase in T_{eff} with M is stronger than observed. For the relation between L and T_{eff} we find $L \propto T_{eff}^{5.2/0.96} = T_{eff}^{5.42}$ or d log L/d log $T_{eff} = 5.42$. For $\alpha = 0.5$ and $\beta = 2.5$ we find $T_{eff} \propto M^{0.72}$ and $L \propto T_{eff}^{5.97}$. In Figs. 10.1–10.3 we compare all these relations with the observed values for main sequence B, A and F stars. The agreement with observation is much better for $\alpha = 0.5$ and $\beta = 2.5$ than for Kramers' opacity law, which is apparently not a good approximation. We realize the importance of good values for κ.

In this simplified discussion, of course, wc do not recover exactly the observed relations but we can find qualitative agreement with observation and can make semi-quantitative comparisons of stars with different masses and chemical compositions.

For stars in which the proton–proton chain is the main energy source, like our sun, we have $\nu = 4$ and we find from equation (10.2) for Kramers' opacity law that

$$R \propto M^{0.5/6.5} = M^{1/13} \tag{10.5}$$

while for $\alpha = 0.5$ and $\beta = 2.5$ we find

$$R \propto M^{1/6} \tag{10.5a}$$

The radius depends only very weakly on the mass. Using equation (10.4) again we find now for Kramers' opacity

$$L \propto \frac{M^{5.5}}{M^{1/26}} \approx M^{5.46} \tag{10.6}$$

and for $\alpha = 0.5$ and $\beta = 2.5$

$$L \propto M^{4.83} \tag{10.6a}$$

We find a somewhat steeper mass–luminosity relation than for the hotter stars, but not much steeper. The radius again enters only very little into this. The mass–luminosity relation is determined mainly by the energy transport.

For T_{eff} we find with Kramers' opacity

$$T_{\text{eff}}^4 \propto \frac{L}{R^2} \propto \frac{M^{5.46}}{M^{1/6.5}} = M^{5.21} \quad \text{or} \quad T_{\text{eff}} \propto M^{1.30} \tag{10.7}$$

and for $\alpha = 0.5$ and $\beta = 2.5$

$$T_{\text{eff}} \propto M^{1.12} \tag{10.7a}$$

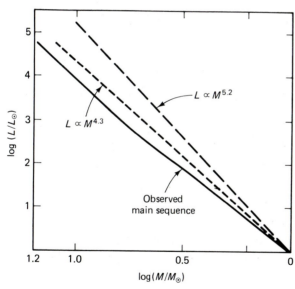

Fig. 10.1. A comparison of empirical (solid line) and theoretical (dashed lines) mass–luminosity relations for homologous stars in radiative equilibrium and CNO energy generation. For Kramers' opacity law we find $L \propto M^{5.2}$ (long dashes). For $\kappa = \kappa_0 \rho^{0.5} T^{-2.5}$ we find $L \propto M^{4.3}$ (short dashes). For the latter approximation the agreement between theoretical and observed relation is much better than for Kramers' opacity law.

In Fig. 10.4 we compare the empirical mass–luminosity relation for the F, G and K stars with this mass–luminosity relation for the idealized homologous, radiative equilibrium models. The overall trend is verified. For the G and K stars of course the outer convection zones become increasingly important.

For the relation between L and T_{eff} we derive with Kramers' opacity law

$$L \propto T_{eff}^{5.46/1.30} = T_{eff}^{4.2} \tag{10.8}$$

and with $\alpha = 0.5, \beta = 2.5$

$$L \propto T_{eff}^{4.3} \tag{10.8a}$$

Both relations are shown in Fig. 10.5. The match with the observations is poor for $T_{eff} > 6000$ K, while for lower T_{eff} it is rather good even though the outer convection zones become important for the energy transport. For a qualitative comparison of stars with different chemical compositions but the same mass we may therefore use these relations with $\beta = 2.5$ and $\alpha = 0.5$.

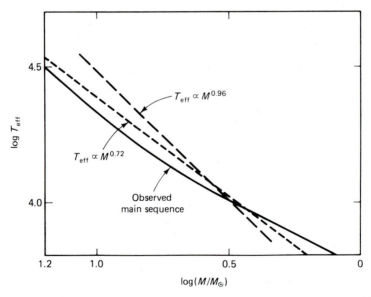

Fig. 10.2. A comparison of the empirical (solid line) and the theoretical (dashed lines) relations between T_{eff} and mass for homologous stars in radiative equilibrium and the CNO cycle as energy source. For Kramers' opacity law we find $T_{eff} \propto M^{0.96}$ (long dashes) and for $\kappa = \kappa_0 \rho^{0.5} T^{-2.5}$ we obtain $T_{eff} \propto M^{0.72}$ (short dashes), which agrees better with the observations.

10.2 Dependence of stellar parameters on the mean atomic weight or evolution of mixed stars

We are now going to discuss the dependence of the main sequence position on the chemical abundances. We will first study the dependence on the mean atomic weight μ, which is mainly determined by the hydrogen and helium abundances. Since hydrogen is converted by nuclear reactions into helium the helium abundance increases during stellar evolution. This leads to an increase in the overall helium abundance *if* the stars remain well mixed during the evolution. The discussion in this section therefore refers to the evolutionary changes of stellar parameters for *mixed stars*. After this we recognize that real stars are generally not well mixed.

The mean atomic weight μ appears explicitly in the constants C and D. The He abundance also enters somewhat into the κ_0, mainly due to the number of free electrons. In the following discussion we omit the changes of κ_0 in the opacity law. This means we cannot extend the derivations to pure He stars. For He stars we also have to consider different nuclear

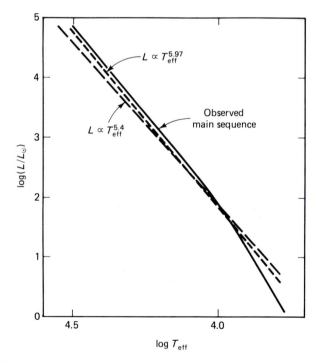

Fig. 10.3. A comparison of the luminosity–T_{eff} relations as observed (solid line) and as derived theoretically (dashed lines) for homologous stars in radiative equilibrium and the CNO cycle as energy source. For Kramers' opacity law we find $L \propto T_{\mathrm{eff}}^{5.4}$ (long dashes) and for $\kappa = \kappa_0 \rho^{0.5} T^{-2.5}$ we obtain $L \propto T_{\mathrm{eff}}^{5.97}$ (short dashes).

reactions. The discussions here therefore refer only to moderate increases in He abundances, say by factors of 2 or 3.

We again multiply the expressions for the constants C and D in order to eliminate the luminosity L, but we now focus our attention on the dependence on μ, i.e. comparing stars of a *given mass* but with different μ. We use $\alpha = 0.5$ and $\beta = 2.5$. We have

$$CD = \text{const.} \frac{\mu^\nu M^{\nu-1-\beta+\alpha}}{\mu^{\beta+4} R^{\nu+3-\beta+3\alpha}} = \mu^{\nu-6.5} R^{-(\nu+2)} M^{\nu-3} \qquad (10.9)$$

For constant mass we obtain

$$R^{\nu+2} \propto \mu^{\nu-6.5} \qquad (10.10)$$

With $\nu = 16$ for the upper part of the main sequence where evolution goes fastest this yields

$$R \propto \mu^{9.5/18} \quad \text{or roughly} \quad R \propto \mu^{1/2} \qquad (10.11)$$

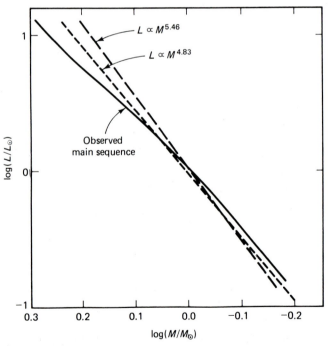

Fig. 10.4. A comparison between empirical (solid line) and theoretical (dashed lines) mass–luminosity relation for low mass homologous stars in radiative equilibrium with the proton–proton chain as energy source. For Kramers' opacity law we find $L \propto M^{5.46}$ which is too steep. For $\kappa = \kappa_0 \rho^{0.5} T^{-2.5}$ a relation $L \propto M^{4.89}$ is found, which shows better agreement with the empirical relation.

From equation (9.17) we then derive

$$L \propto \mu^{6.5}/R \propto \mu^6 \qquad (10.12)$$

The luminosity increases steeply with increasing mean molecular weight!
A higher central temperature is required in order to increase the central
pressure enough to balance the weight of the overlying material. The
radius also depends on the mean atomic weight. The dependence of the
radius on μ does however change for a different energy generation
mechanism. For stars on the lower main sequence when the proton–
proton chain is supplying the nuclear energy, i.e. when $\nu = 4$, we find

$$R \propto \mu^{-2.5/6} \quad \text{or roughly} \quad R \propto \mu^{-0.4} \qquad (10.13)$$

While for the upper main sequence stars the radius increases for
increasing μ we find for the lower main sequence stars a decrease in R with
increasing μ. In this case the luminosity increases even more steeply with

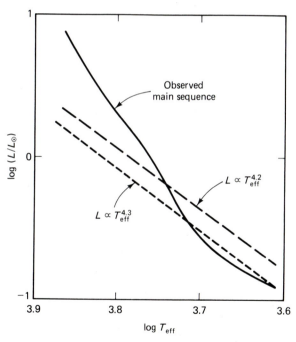

Fig. 10.5. A comparison between the observed main sequence (solid line) and the
theoretical relations for low mass homologous stars in radiative equilibrium and with the
proton–proton chain as energy source. For Kramers' opacity law a relation $L \propto T_{\text{eff}}^{4.2}$ is
found (long dashes). For $\kappa = \kappa_0 \rho^{0.5} T^{-2.5}$ a relation $L \propto T_{\text{eff}}^{4.3}$ is derived (short dashes).
For low temperatures good agreement is found between observed and theoretical
relations.

increasing μ; we find for stars in which the proton–proton chain provides the energy

$$L \propto \mu^{6.5}/R \propto \mu^{6.9} \quad \text{roughly for} \quad \nu = 4 \tag{10.14}$$

Knowing the luminosity and radius we can also estimate the dependence of T_{eff} on μ, though we know that this will be less accurate. We find for $\nu = 16$

$$T_{\text{eff}}^4 \propto \frac{L}{R^2} \propto \frac{\mu^6}{\mu} = \mu^5 \quad \text{and} \quad T_{\text{eff}} \propto \mu^{1.25} \quad \text{approximately} \tag{10.15}$$

If $\nu = 4$ we find

$$T_{\text{eff}}^4 \propto \mu^{7.76} \quad \text{and} \quad T_{\text{eff}} \propto \mu^{1.94} \quad \text{approximately} \tag{10.16}$$

In any case the stars become much more luminous and their effective temperatures increase. They evolve upwards and to the left in the HR diagram. The question is do they evolve above or below the main sequence of the hydrogen rich stars? In order to see this we have to compare the $L(T_{\text{eff}})$ relation for a given mass but increasing μ with the relation between L and T_{eff} for the main sequence stars, i.e. stars of given μ but different M. If, for increasing μ, T_{eff} increases more steeply with L than for main sequence stars, the mixed stars will evolve below the main sequence. If T_{eff} increases more slowly with L than for the main sequence they will evolve above the main sequence. For $\nu = 16$ we find from equations (10.12) and (10.15) that for a given mass,

$$T_{\text{eff}} \propto \mu^{1.25} \propto L^{1.25/6} = L^{0.208} \quad \text{or} \quad L \propto T_{\text{eff}}^{4.8} \tag{10.17}$$

and for $\nu = 4$ we have

$$T_{\text{eff}} \propto \mu^{1.9} \propto L^{1.9/6.9} = L^{0.28} \quad \text{or} \quad L \propto T_{\text{eff}}^{3.63} \tag{10.18}$$

These relations have to be compared with the main sequence relations obtained in the same way looking at stars with different masses but the same μ. We had found for $\nu = 16$ (see the preceding section)

$$T_{\text{eff}} \propto L^{0.72/4.28} = L^{0.17} \quad \text{or} \quad L \propto T_{\text{eff}}^{5.94} \tag{10.19}$$

For $\nu = 4$ we derived for the main sequence

$$T_{\text{eff}} \propto L^{0.23} \quad \text{or} \quad L \propto T_{\text{eff}}^{4.3} \tag{10.20}$$

In both cases we find that, for increasing values of μ, T_{eff} increases more steeply with increasing L than on the main sequence which means the stars evolve to the left of the main sequence, as shown in Fig. 10.6. **Were the** stars to remain mixed during their evolution they would evolve up to larger *L* but evolve below the well-known main sequence.

Calculations determining the position of the *pure helium star* main sequence, taking into account the change in the absorption coefficients and the change in the energy generation (which has to be by the triple-alpha reaction because there is no hydrogen for the other reactions), find that the main sequence for pure helium stars is far below the hydrogen star main sequence even though for a given mass the luminosity of the helium star is larger but its T_{eff} is also much larger. In Fig. 10.7 we compare the solar composition main sequence with the main sequence of helium stars as calculated by Cox and Salpeter (1964). Helium-rich stars are found to the left of the normal main sequence. For a given mass L and T_{eff} are larger than for hydrogen-rich stars.

10.3 Changes of main sequence position for decreasing heavy element abundances

For lower mass stars, we would like to compare population I and population II stars, i.e., stars with different metal abundances Z. Stars with different abundances of heavy elements are not strictly homologous

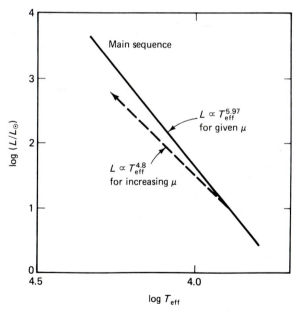

Fig. 10.6. The evolutionary track for a mixed star, i.e. a star with a given mass but increasing μ, is compared with the main sequence, i.e. with the positions of stars with given μ but different masses. Homologous stars in radiative equilibrium are considered with the CNO cycle as energy source.

because the depth of the outer convection zone depends on Z. Neverthe-less comparing homologous stars still shows us the trend of the differences. Since the heavy elements contribute such a small mass fraction decreasing heavy element abundances will not influence μ. The abundance of the heavy elements enters mainly into the κ_0. In the deep interior the bound-free κ_0 is proportional to Z because only heavy element ions still have electrons bound to them which can absorb photons. Since there are free-free contributions to κ which do not depend on Z the use of the bound-free κ alone will exaggerate the dependence on Z but still give the qualitative trend.

If the energy source is the proton–proton chain as we expect for stars with about one solar mass and less, then ε_0 does not depend on the metal abundance.

We again multiply the constants C and D to eliminate the luminosity. Including the Z dependence of κ_0 this yields

$$CD \propto \frac{ZRM^{\nu+2}}{R^{\nu+3}M^5} \tag{10.21}$$

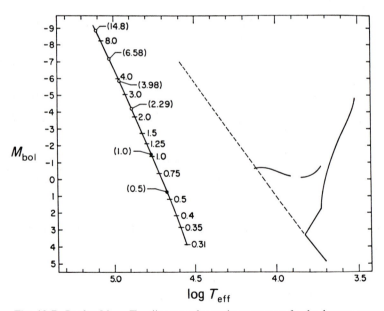

Fig. 10.7. In the M_{bol}, T_{eff} diagram the main sequence for hydrogen stars (dashed line) is compared with the main sequence for helium stars as calculated by Cox and Salpeter. The stellar masses in units of solar masses are given at the appropriate points. (Dots and circles are from less accurate calculations.) Also shown in the lower right is a globular cluster diagram for comparison. From Cox and Salpeter (1964).

With $\nu \approx 4$ we obtain

$$CD \propto ZR^{-6}M \tag{10.22}$$

which gives for constant M

$$R^6 \propto Z \quad \text{or} \quad R \propto Z^{1/6} \tag{10.23}$$

The radius decreases only very little with decreasing Z.

Inserting this into equation (9.17) we find for C with constant mass

$$C \propto ZLR \propto Z^{7/6}L \tag{10.24}$$

For the luminosity we find from equation (10.24) that

$$L \propto Z^{-7/6} \tag{10.25}$$

For stars of a given mass the luminosity increases with decreasing Z! For a given mass the lifetime of a metal-poor star therefore is shorter than for a metal-rich star because it uses its fuel more rapidly. As we said earlier, equation (10.25) exaggerates the Z dependence of L but gives the correct trend: **for smaller Z, the κ decreases and the radiation can escape more easily. The luminosity increases.**

We are now looking at the change of T_{eff}. We obtain

$$T_{\text{eff}}^4 \propto \frac{L}{R^2} \propto \frac{1}{Z^{2/6}Z^{7/6}} \approx Z^{-9/6} \tag{10.26}$$

or

$$T_{\text{eff}} \propto Z^{-0.376} \sim Z^{-1/3} \tag{10.27}$$

The Z dependence is again exaggerated but shows the correct trend. **L and T_{eff} both increase for decreasing Z.**

For the relation between T_{eff} and L we then find

$$L \propto T_{\text{eff}}^{3.1} \quad \text{or} \quad T_{\text{eff}} \propto L^{0.33} \tag{10.28}$$

for stars with changing Z but constant mass.

With the same approximations we found for the luminosity T_{eff} relation along the solar composition main sequence, i.e. for stars with the same Z but different masses (equation 10.8)

$$L \propto T_{\text{eff}}^{4.3} \quad \text{approximately}$$

Again, **the main sequence has a steeper increase of L with increasing T_{eff} than the track in the HR diagram** outlining the change in appearance **for stars of decreasing heavy element abundances,** shown in Fig. 10.8. Stars with a given mass but different Z will have to follow the trend of this track.

The main sequence for metal-poor population II stars lies below the main sequence for solar abundance stars, at least if the helium abundance for these stars is the same as for the population I stars.

We are still not sure about the helium abundance in population II stars because only cool population II stars are bright enough to permit a good spectrum analysis, and cool stars do not show helium lines. The only hot population II stars are the blue horizontal branch stars and they are so faint that up until now no spectrum analysis has been possible. With more sensitive receiving instruments we may hope to have some helium abundance determination for these stars soon, but even then we will still not know whether the helium abundance in these very evolved stars is the same as the one for the main sequence population II stars.

Since helium is constantly manufactured in hot stars and during their final evolution stages is mixed in with the interstellar medium we might expect the more recently formed population I stars to have a higher helium abundance than the old population II stars. If the metal-poor population II stars do indeed have a smaller helium abundance than population I stars,

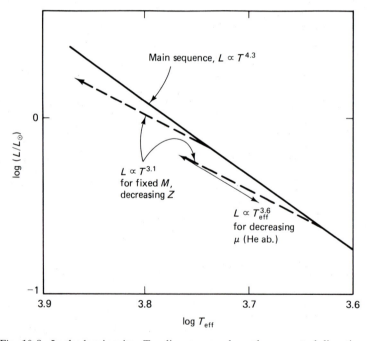

Fig. 10.8. In the luminosity, T_{eff} diagram we show the expected direction of change for decreasing abundances of heavy elements Z (long dashes) and the direction of change in position for decreasing μ due to possibly lower helium abundance (solid arrow). Also shown is the main sequence (solid line). Homologous stars in radiative equilibrium are compared.

we have to consider the influence of the lower helium abundance simultaneously with the *decreasing* metal abundances. As we saw in the preceding section, a decreasing helium abundance moves the stars to lower luminosities and lower T_{eff}, i.e. in the opposite direction to the trend we discussed for increasing helium abundances. A small decrease in helium abundance would move the population II stars back along the line $L \propto T_{\text{eff}}^{3.57}$, i.e. would make the main sequence for metal-poor stars agree more with the one for solar abundance stars. For the same L or T_{eff} a population II star would still have a smaller mass (see Fig. 10.8). Because of the uncertainty of the helium abundance for population II stars we are still not quite sure about their masses. Their absolute magnitudes are also not known accurately. We have to rely on uncertain trigonometric parallaxes for some relatively close population II dwarfs. It appears, however, that the population II main sequence does lie below the population I main sequence, as seen in Fig. 10.9.

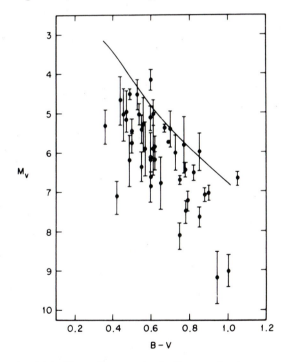

Fig. 10.9. The color magnitude diagram for the nearest population II stars, for which trigonometric parallaxes can be measured. The vertical lines indicate error bars. The solid line shows the main sequence for stars with solar element abundances. When compared at the same B − V the main sequence for very metal-poor stars appears to be about two magnitudes below the solar abundance main sequence. When compared at equal T_{eff} the difference is smaller, because for a given T_{eff} the B − V decreases for metal-deficient stars. From Sandage (1986).

In Fig. 10.10 we show theoretical sequences calculated for stars with different abundances of heavy elements and helium, calculated to the best of our present knowledge with the inclusion of all complications.

10.4   Homologous contracting stars in radiative equilibrium

10.4.1 T_{eff}–luminosity relations

During the early evolutionary phases young stars are still rather cool in their interiors. We saw (equation 2.19) that hydrostatic equilibrium requires their thermal energy E_{thermal} to be half the amount of the released gravitational energy $E_{\text{grav}} \approx GM^2/R$, i.e.

$$E_{\text{thermal}} = \frac{1}{2}\frac{GM^2}{R} \tag{10.29}$$

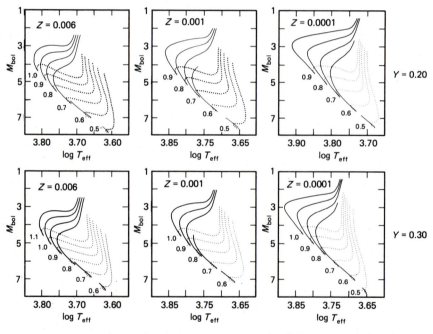

Fig. 10.10. In the M_{bol}, $\log T_{\text{eff}}$ diagram main sequences for different metal abundances Z and helium abundances are shown. A helium abundance $Y = 0.30$ was assumed for the lower panel and $Y = 0.20$ for the upper panel. Stellar masses (in units of the solar mass) are given at the point of arrival on the main sequence. These points outline the so-called zero age main sequence (ZAMS). The shift of stellar positions for a given mass to higher T_{eff} and L for increasing Y and decreasing Z is obvious. Note the change in the T_{eff} scale for the different panels. The dotted lines outline the evolution during the contraction phases (see also chapter 11). From VandenBerg and Bell (1985).

For large radii the gravitational energy released is still small and therefore the internal temperatures are still low, not high enough for nuclear reactions to take place. The quantity D derived from the equations for nuclear energy generation is therefore not applicable for the discussion of contracting stars. Fortunately we can obtain all the information we need from the constant C alone. During contraction of a star the mass remains constant (except perhaps for the very early phases). From the condition $C = $ constant we then find for $\kappa_g = \kappa_0 \rho^{0.5} T^{-2.5}$

$$L \propto R^{-1} \qquad (10.30)$$

With decreasing radius the luminosity increases slowly. From the relation

$$T_{\text{eff}}^4 \propto L/R^2$$

we find with $L \propto R^{-1}$ that

$$T_{\text{eff}}^4 \propto R^{-3} \propto L^3 \qquad (10.31)$$

The effective temperature increases rapidly with decreasing radius, and very rapidly with increasing luminosity. We find $T_{\text{eff}} \propto L^{3/4}$. During contraction the star evolves almost horizontally through the HR diagram if it is in radiative equilibrium (see Fig. 10.10).

In Section 8.1 we estimated how long the Sun could live on its gravitational energy assuming that it had constant luminosity during the contraction. We now see that this assumption is not very wrong. The effective temperature changes quickly with a small increase in luminosity. An increase in M_{bol} by 0.75 correlates with an increase in T_{eff} from say 4000 K to 6700 K (see Fig. 10.10), depending on Z and M.

Do stars evolve from a cool interstellar cloud by slowly contracting along the relation $L \propto R^{-1}$? (More accurate calculations, taking into account the accurate $\kappa(T, P_g)$, show a dependence $L \propto R^{-0.75}$.) The answer is no. Observations show that this is not the case. In star-forming regions of the sky where we find young, luminous stars we do not find *very* cool, low luminosity stars which would fall along this track. The track derived here for contracting stars only holds for the higher temperature part, i.e. $T \gtrsim 5000$ K as seen in Fig. 10.10, which means only for stars with no deep convection zones. So far we have neglected convection in our discussions of contracting stars and of main sequence stars. For cool stars the outer convection zones become very important. The neglect of these zones leads to the failure of our predictions for cool, contracting stars as we shall see soon.

10.4.2 Energy release in a contracting star

In order to calculate the temperature stratification, according to equations (3.9) and (9.1), we have to calculate the gravitational energy release as a function of depth.

The gravitational energy release is due to the shrinking of the star. For the whole star it can be expressed as

$$-\frac{\mathrm{d}E_G}{\mathrm{d}t} = \frac{\mathrm{d}}{\mathrm{d}t} \int_0^R \frac{GM_r}{r} \rho 4\pi r^2 \, \mathrm{d}r \propto \frac{\mathrm{d}}{\mathrm{d}t} \frac{GM^2}{R} \tag{10.32}$$

where t stands for time.

The question remains how this energy release is distributed over the star. How much is liberated at any given shell with radius r within the star?

At any given point in the star we can quite generally say that the energy content per gram of material is altered by three effects: (a) by the work done on the volume against the gas pressure, given by $P_g \, \mathrm{d}V$, where V is the specific volume, i.e. the volume which contains 1 g of material; (b) the energy content is increased by the nuclear energy generation ε_n; (c) it is decreased by the difference of energy flux leaving the volume of gas and the flux entering the volume. Per cm^3, the latter amount is given by $\mathrm{d}\pi F/\mathrm{d}r$ and per gram the amount must then be $(\mathrm{d}\pi F/\mathrm{d}r)\rho^{-1}$. The energy equilibrium requires

$$\frac{\mathrm{d}}{\mathrm{d}t}\left(\frac{3}{2}\frac{R_g}{\mu}T\right) = -P_g \frac{\mathrm{d}}{\mathrm{d}t}\left(\frac{1}{\rho}\right) + \varepsilon_n - \frac{1}{4\pi r^2}\frac{1}{\rho}\frac{\mathrm{d}L}{\mathrm{d}r} \tag{10.33}$$

The left-hand side describes the thermal energy content per gram of material (if only kinetic energy needs to be considered). On the right-hand side we have replaced πF by $L/4\pi r^2$. With $P_g = \rho R_g T/\mu$ the left-hand side gives

$$\frac{\mathrm{d}}{\mathrm{d}t}\left(\frac{3}{2}\frac{P_g}{\rho}\right) = -\frac{3}{2}\frac{P_g}{\rho^2}\frac{\mathrm{d}\rho}{\mathrm{d}t} + \frac{3}{2}\frac{1}{\rho}\frac{\mathrm{d}P_g}{\mathrm{d}t} \tag{10.34}$$

This can be combined with the first term on the right-hand side to give

$$-\frac{5}{2}\frac{P_g}{\rho^2}\frac{\mathrm{d}\rho}{\mathrm{d}t} + \frac{3}{2}\frac{1}{\rho}\frac{\mathrm{d}P_g}{\mathrm{d}t} = \frac{3}{2}\rho^{2/3}\frac{\mathrm{d}}{\mathrm{d}t}\left(\frac{P_g}{\rho^{5/3}}\right) \tag{10.35}$$

as can be verified easily. The energy equation (10.33) becomes

$$\frac{3}{2}\rho^{2/3}\frac{\mathrm{d}}{\mathrm{d}t}\left(\frac{P_g}{\rho^{5/3}}\right) = \varepsilon_n - \frac{1}{4\pi r^2}\frac{1}{\rho}\frac{\mathrm{d}L}{\mathrm{d}r} \tag{10.36}$$

How does this relate to the change in gravitational energy as described by equation (10.32)? This is, of course, contained in the left-hand side which includes the term $P_g \, dV$. The gravitational force does this work when the star contracts. For homologous stars we can see how this term is related to the change in radius of the star. For different homologous stars with a given mass we know we must have at any given point r/R_0

$$\rho = \rho_0 \left(\frac{R}{R_0}\right)^{-3}, \quad P_g = P_{g0} \left(\frac{R}{R_0}\right)^{-4} \quad \text{and} \quad T = T_0 \left(\frac{R}{R_0}\right)^{-1} \quad (10.37)$$

where ρ_0, P_{g0}, T_0 are the values of ρ, P_g and T for the star with radius R_0.

We can now write the left-hand side of equation (10.36) in the form

$$\frac{3}{2} \rho^{2/3} \frac{d}{dt} \left(\frac{P_g}{\rho^{5/3}}\right) = \frac{3}{2} \rho^{2/3} \frac{P_g}{\rho^{5/3}} \frac{d}{dt} \left(\ln \frac{P_g}{\rho^{5/3}}\right)$$

$$= \frac{3}{2} \frac{P_g}{\rho} \left[\frac{d}{dt} \ln (R^{-4}) - \frac{5}{3} \frac{d}{dt} \ln (R^{-3})\right]$$

$$= \frac{3}{2} \frac{P_g}{\rho} \frac{d \ln R}{dt} \quad (10.38)$$

which clearly shows the relation to the changes of the radius.

11

Influence of convection zones on stellar structure

11.1   Changes in radius, luminosity and effective temperature

In the previous chapter we considered only model stars in radiative equilibrium. We pointed out several mismatches between these models with real stars and attributed them in part to the influence of convection zones. Convection zones change stellar structure in two main ways:

(a) The radius of the star becomes smaller.

(b) The energy transport through the outer convection zones with the large absorption coefficients becomes easier due to the additional convective energy transport, so that the temperature gradient becomes smaller in comparison with radiative equilibrium. This may lead to an increased luminosity and T_{eff} as well as energy generation.

If energy transport outwards due to convection is increased the star would tend to lose more energy than is generated, and so would tend to cool off. However, this does not actually happen, because it would reduce the internal gas pressure and the gravitational pull would then exceed the pressure force. The star actually contracts, the stellar interior temperature increases, thereby increasing the energy generation $\varepsilon \propto T^{\nu}$. With the larger energy generation the star is then able to balance the larger energy loss. The star is again in thermal equilibrium but with a smaller radius and a larger luminosity, which means with a larger effective temperature. As compared to radiative equilibrium the star moves to the left and up in the HR diagram (see Fig. 11.1). **Convection decreases the equilibrium value for the radius.**

We can also look at it from another angle by the following 'thought experiment'. We imagine a star with a given T_{eff} and radiative equilibrium $T(r)$. This star is however unstable to convection in the outer convection zones. This means in these layers the radiative equilibrium temperature gradient is steeper than the adiabatic one. We plot the radiative equilibrium temperature stratification schematically in Fig. 11.2. Since the star is

130

in hydrostatic equilibrium the interior pressure just balances the weight of the overlying material. (This is the kind of star which we have dealt with so far when discussing cool homologous stars in radiative equilibrium.) However, with $\nabla_{\text{rad}} > \nabla_{\text{ad}}$ the radiative equilibrium is unstable. If we introduce an infinitely small perturbation convection will start. In the convection zones the radiative energy transport is reduced by the amount of the convective energy transport, which means with convection the temperature gradient and thereby the temperature is reduced everywhere below the upper boundary of the convection zone. The central temperature and pressure would therefore also be reduced. The star must contract and the thermal energy $E_{\text{therm}} = -\frac{1}{2}E_{\text{grav}}$ must increase. This must continue until enough gravitational energy is released to heat the star everywhere to temperatures which are on average, and especially in the center, higher than they were originally in the radiative equilibrium star. Because of the higher central temperature and density nuclear energy generation increases, leading to a larger luminosity and effective temperature of the star.

All these effects become most noticeable if the temperature gradient is strongly reduced with the onset of convection, i.e. if the convective energy transport is very efficient. The observed change in stellar radius as compared with radiative equilibrium models is therefore a measure for the efficiency of the convective energy transport, which depends on the degree of convective instability and also on the size of the characteristic traveling length l. The observed stellar radii can therefore be used to determine the characteristic length l. Observed stellar radii, when compared with

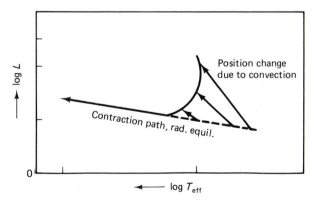

Fig. 11.1. Shows schematically for cool stars the change of position in the HR diagram due to increasing efficiencies of convective energy transport with decreasing T_{eff}. For more efficient energy transport a star of a given mass has to have a higher luminosity and effective temperature. Its radius decreases.

modern calculations, indicate a value of $l/H \approx 1.5$ for lower main sequence stars. We do not know, however, whether the ratio of l/H is the same for all kinds of stars. We remember that the Li^7 abundances observed in the solar photosphere lead to an estimate $l/H \approx 1$ as discussed in Chapter 7.

11.2   The Hayashi line

In this section we will talk about completely convective stars. We saw in Chapter 10 that for the convective regions the constant E in equation (9.15) takes the place of the constant C in equation (9.14) (which determines the relation between M, L and R for stars in radiative equilibrium). This constant E is related to the stellar parameters in the following way (see equation 9.19)

$$E = 4\pi K G^{3/2} (\mu/R_g)^{5/2} M^{1/2} R^{3/2} \tag{11.1}$$

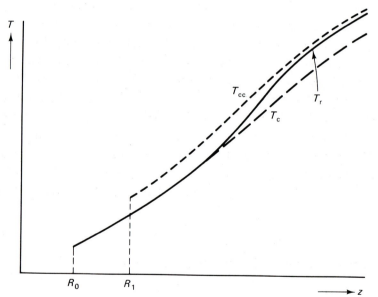

Fig. 11.2. The depth dependence of the temperature T_r for a star in radiative equilibrium is shown schematically (solid line). In a star which is convectively unstable the onset of convection reduces the temperature gradient in the convection zone, as indicated by T_c (long dashed line). The central temperature would then also be reduced. This would decrease P_g (center) below the equilibrium value. The star must shrink from the radius R_0 to the radius R_1 and heat up to the temperatures shown as T_{cc} (short dashes).

where the constant K determines the adiabat which is followed by the temperature stratification:

$$T^{5/2} = P_g/K \quad \text{with} \quad \frac{d \ln T}{d \ln P_g} = 0.4 \qquad (11.2)$$

for adiabatic stratification with $\gamma = \frac{5}{3}$. For the convective layers the constant E is determined by the inner and outer boundary conditions. If inner and outer radiative equilibrium zones are considered E is determined by the conditions that at the boundaries of the convection zone(s) temperature and pressure and their gradients have to match the values of the adjacent radiative equilibrium layers. It is found that completely convective stars have the largest value of E, namely $E_0 = 45.48$ (Hayashi, Hoshi and Sugimoto 1962). If there is a radiative equilibrium zone below the convection zone then these radiative zones can only be matched with a convective zone having a smaller value of E.

How do we determine the positions of these stars in the HR diagram? We have to use the constant E instead of the constant C which we used for radiative equilibrium stars. For a given $E = E_0$ and for a given mass, equation (11.1) determines a relation $K = K(R)$. In order to determine both K and R, we need another equation. This equation comes from the boundary condition at the top of the adiabatically stratified region, which is, of course, the bottom of the stellar atmosphere. The stellar *atmosphere* is determined by T_{eff} and the gravitational acceleration $g = GM/R^2$. For a given mass $g = g(R)$. For a given $g(R)$ we can integrate the hydrostatic equilibrium equation down to the bottom of the atmosphere where the optical depth is $\tau = \tau_1$ (see for instance equation 5.18). This gives us $P_g(\tau_1)$. For a given T_{eff} we can also calculate the temperature for this same optical depth τ_1 and obtain $T(\tau_1)$ according to equations (5.20) and (5.21).

With $T(\tau_1)$ known and $P_g(\tau_1)$ known, both depending on R and T_{eff}, we determine

$$K = P_g(\tau_1)/T(\tau_1)^{5/2} = K(R, T_{\text{eff}}) = K(R, L) \qquad (11.3)$$

where we have eliminated T_{eff} by means of the relation

$$T_{\text{eff}}^4 = \frac{L}{\sigma 4 \pi R^2} \qquad (11.4)$$

The relation (11.3) provides the second equation to determine K and R for a given M. It contains, however, as a new parameter the luminosity L. For a given set of M and E we thus find a one-dimensional sequence of solutions for K and R as a function of L.

For luminosities $L \leqslant L_\odot$ Hayashi, Hoshi and Sugimoto (1962) succeeded in giving an approximate analytical relation between the stellar parameters, namely

$$\log \frac{L}{L_\odot} = 0.272 - 1.835 \log \frac{M}{M_\odot} + 9.17(\log T_{\mathrm{eff}} - 3.70)$$

$$+ 2.27 \log \frac{E_0}{40} + 0.699(\log \kappa_0(Z) + 15.58) \qquad (11.5)$$

It is of special importance to note the very strong increase of L with T_{eff}. Changing T_{eff} by 4 per cent changes L by roughly a factor of 10 and M_{bol} by about -2.5. **For a given value of E, stars of a given mass lie essentially on a vertical line in the L versus T_{eff} diagram as seen in Fig. 11.3.** We find

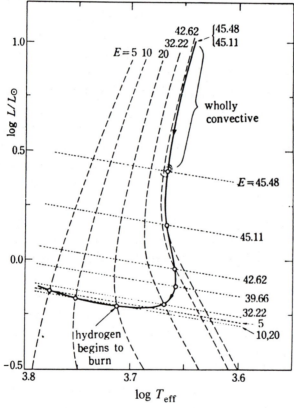

Fig. 11.3. Stars for a given constant E lie along a nearly vertical line in the T_{eff}, luminosity diagram (dashed lines). For decreasing values of E the T_{eff} increases for a given value of L. The nearly horizontal dotted lines show the combinations of L and T_{eff} permitted by the atmospheric boundary conditions for a given E. The intersection with the vertical $E = $ constant line gives the actual position of the star ($\circ$). The thick solid line outlines the actual evolutionary track of a contracting star which slowly develops a growing radiative equilibrium core. From Hayashi, Hoshi and Sugimoto (1962).

(d log L/d log T_{eff}) = 9.17 to be compared with the value obtained for the main sequence (d log L/d log T_{eff}) $\approx$ 5. The almost vertical line in the HR diagram on which fully convective stars of a given mass, i.e. stars with $E = 45.58$, are found is called the *Hayashi line* after the astronomer who first derived the relations for fully convective stars and discussed the consequences.

As we said earlier the largest value of E corresponds to completely convective stars. Smaller values of E lead to higher T_{eff} for a given L, according to equation (11.5). They correspond to stars with an interior zone in radiative equilibrium. To know which value of E is actually applicable we have to test the fully adiabatic star for convective instability in the interior. If $\nabla_r < \nabla_{\text{ad}}$ in the core then a model with a smaller E is applicable. These stars then have larger T_{eff}. The nearly vertical lines in Fig. 11.3 combine stars with a given E, i.e. homologous stars, with similar extent of the radiative equilibrium cores. The intersection of these lines with the horizontal $E = $ constant line, as obtained from the outer boundary condition, yields the possible stellar models satisfying all interior and surface constraints. The actual evolutionary track followed by a contracting star is then given by the thick line following the Hayashi track in the beginning. The star slowly develops a growing core in radiative equilibrium and finally with increasing T_{eff} becomes a star which is almost completely in radiative equilibrium following the track which we calculated for homologous contracting stars in radiative equilibrium, until it reaches the main sequence when hydrogen burning starts.

Since stars with radiative equilibrium zones have higher T_{eff} than fully convective stars there can be no stars with T_{eff} lower than for the completely convective stars. **The Hayashi line gives a lower limit for the T_{eff} of stars in hydrostatic equilibrium.**

Equation (11.5) shows that for a given L the T_{eff} on the Hayashi line increases with increasing mass, but only slightly. If M increases by a factor of 10 the T_{eff} increases by 50 per cent for constant L. In Fig. 11.4 we show the Hayashi lines for different values of M.

For increasing abundances of heavy elements κ_0 increases and for a given L and M the T_{eff} must decrease. From equation (11.5) we estimate that

$$9.17\Delta \log T_{\text{eff}} = -0.699\Delta \log \kappa_0(Z)$$

for a given L and M. For $\Delta \log \kappa_0(Z) = \Delta \log Z$ in the cool stellar atmospheres we find $\Delta \log T_{\text{eff}} = -0.076\Delta \log Z$. For a change in Z by a factor 100, T_{eff} increases by 40 per cent if $\kappa_0 \propto Z$.

11.3   Physical interpretation of the Hayashi line

How can we understand the existence of the Hayashi line and its dependence on mass and chemical composition? Why is there a lower limit for T_{eff} of stars in hydrostatic equilibrium? We start our discussion by first considering the stellar interior. The central temperature of the star in hydrostatic equilibrium is determined by the mass which means essentially by the central regions containing most of the mass. It is independent of the atmospheric layers. On the other hand κ_0 and L are introduced into the equation (11.5) for the luminosity only by means of the atmospheric boundary conditions. This tells us that it is mainly the atmosphere which determines the relation between L, M and T_{eff}. We now look at the temperature stratification in the star starting from the given central temperature, as shown in the schematic Fig. 11.5. For a fully convective star the temperature decreases outwards adiabatically up to the upper

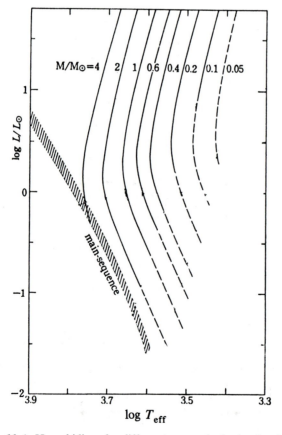

Fig. 11.4. Hayashi lines for different masses in the luminosity, T_{eff} diagram. From Hayashi, Hoshi and Sugimoto (1962).

boundary of the region with efficient convection. For higher layers we rapidly approach radiative equilibrium. For our schematic discussion we assume that out to the layer with $P_g = P_{g1}$ the stratification is adiabatic and that for $P_g < P_{g1}$ the stratification follows the one for radiative equilibrium with $\nabla_r < \nabla_{ad}$. In these high layers the temperature gradient is inversely proportional to the absorption coefficient. For cool stars the continuous absorption in the atmospheres is due to the H^- ion (see Volume 2). For very cool stars there are few free electrons to form the H^- ion. (Some heavy atoms have to be ionized to provide the electrons.) So κ is very small. The radiative equilibrium temperature gradient Δ_{rad} is therefore very flat and the gas pressure P_{g0} at $\bar{\tau} = \frac{2}{3}$ is reached at a fairly high temperature $T_{eff,1}$. With such **low values of κ** and the corresponding small temperature gradient **very low effective temperatures cannot be reached.** For larger heavy element abundances the κ_0 increases because more free electrons become available to form H^-, the temperature gradient steepens and somewhat lower surface temperatures $T_{eff,2}$ can be reached.

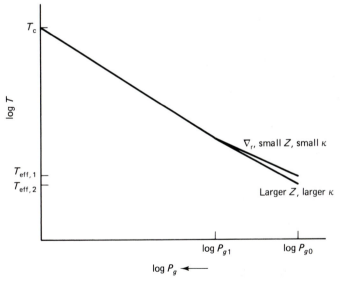

Fig. 11.5. Shows schematically the temperature stratification in a cool star. The central temperature of a star, T_c, is determined by its mass. In a fully convective star the temperature decrease outwards follows ∇_{ad}. The top layer of the star with $P_g > P_{g1}$ is in radiative equilibrium. The temperature gradient ∇_{rad} is proportional to the absorption coefficient κ. At the layer with $\bar{\tau} = \frac{2}{3}$ and with $P_g = P_{g0}$ a temperature $T = T_{eff}$ is reached which depends on ∇_{rad}. The larger κ the steeper ∇_{rad} and the lower is T_{eff}. For cool stars κ becomes very small and T_{eff} cannot become very low. For larger κ, which means for a larger metal abundance Z, T_{eff} can become lower than for the low Z. The lowest possible value is determined by the stellar mass, determining T_c, and by κ.

We now look at stars of different masses (see Fig. 11.6). For the larger mass star 1 the central temperature T_{c1} must be larger than for lower mass star 2 because of the larger weight of the overlying material. In both stars the temperature decreases adiabatically outwards until the upper boundary of the layer with efficient convective energy transport is reached at a region with $P_g = P_{g1}$ or P_{g2}. (Because of the lower T_{eff} the P_{g2} may be somewhat larger than P_{g1}.) For higher layers we have essentially radiative equilibrium. Since at the layer with $P_g = P_{g1}$ the temperature is higher in the higher mass star the layer with $\bar{\tau} = \frac{2}{3}$ is also reached with a higher temperature. The higher mass star has a higher T_{eff}.

We now compare a star which has a radiative equilibrium zone in the center with a completely adiabatically stratified star. Star 1 with the radiative equilibrium core has in the interior a flatter temperature stratification than the adiabatically stratified fully convective star 2. The situation is shown qualitatively in Fig. 11.7. In star 1 the adiabatic, convective region starts at the layer with $P_g = P_{g2}$; the temperature decreases adiabatically outwards. At the upper boundary of the adiabatic region a higher temperature is found than in the fully convective case, and the star has a higher temperature at $\bar{\tau} = \frac{2}{3}$, which means it has a higher T_{eff}. Stars which are partially in radiative equilibrium have a higher T_{eff} than fully convective stars.

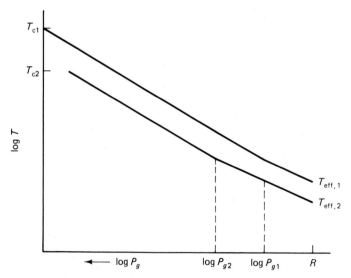

Fig. 11.6. Compares schematically the temperature stratifications as a function of pressure for two stars of different masses. For the larger mass star the central pressure and temperature are higher. A higher T_{eff} is reached at the surface for the higher mass star.

This discussion is of course very schematic, complications are not considered, but it demonstrates the main effects which determine the difference in structure of fully and partially convective stars.

11.4   Stars on the cool side of the Hayashi line

As we pointed out earlier for stars which have radiative equilibrium zones T_{eff} must increase in comparison with fully convective stars on the Hayashi line. There can then be no stars on the cool side of the Hayashi line, at least no stars in hydrostatic equilibrium. **The coolest stars are fully convective and must lie on the Hayashi line.**

There are, however, a few objects observed on the cool side of the Hayashi line with rather low luminosities and temperatures around a few hundred degrees. How can this be? Several explanations are possible:

1. These stars are not in hydrostatic equilibrium.

2. There may be a dust cloud around a hotter star. The stellar radiation proper is then obscured by the dust which may be heated to

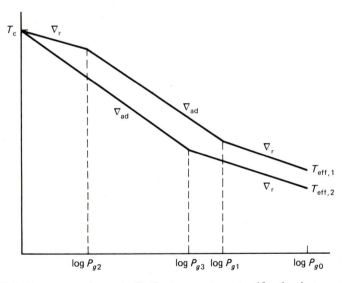

Fig. 11.7. Compares schematically the temperature stratification in two stars, one of which has a central radiative equilibrium core, the other being convective all the way to the center. For stars with a central radiative equilibrium zone, i.e. $\nabla_r < \nabla_{ad}$, the temperature decrease for the central region is slower than for the fully convective stars. In the convection zone starting at $P_g = P_{g2}$ the temperature stratification follows ∇_{ad}. The pressure P_g at the upper boundary of the convection zone is reached for a higher temperature T_1, which leads to a higher value of T_{eff} for $P_g = P_{g0}$ at $\tau = \frac{2}{3}$. The more extended the interior radiative equilibrium zone, the higher T_{eff}.

several hundred degrees by the stellar radiation. We actually see the warm dust cloud.

3. The star may have $\nabla_r > \nabla_{ad}$ but convective energy transport may be inhibited by a stellar magnetic field. In this special case we could have a situation where $\nabla > \nabla_{ad}$, which would put the star on the cool side of the Hayashi line.

12

Calculation of stellar models

12.1   Schwarzschild's method

Before we can discuss the detailed structure of the stars on the main sequence we have to outline the methods by which it can be calculated. In Chapter 10 we have compared homologous stars on the main sequence. While we were able to see how temperatures and pressures in the stars change qualitatively with changing mass and chemical composition, we have never calculated what the radius and effective temperature of a star with a given mass really is. In order to do this we need to integrate the basic differential equations, which determine the stellar structure as outlined in Chapter 9. Two methods are in use: *Schwarzschild's method* and *Henyey's method*.

Schwarzschild's method is described in his book on stellar structure and evolution (1958). The basic differential equations are integrated both from the inside out and from the outside in. In the dimensionless form the differential equations for the integration from the outside in contain the unknown constant C (see Chapter 9), for the integration from the inside out the differential equations also contain the unknown constant D. A series of integrations from both sides of the star is performed for different values of these constants. The problem then is to find the correct values for the constants C and D and thereby the correct solutions for the stellar structure. At some fitting point $x_f = (r/R)_f$ we have to fit the exterior and the interior solutions together in order to get the solution for the whole star. At this fitting point we must of course require that pressure and temperature are continuous. In other words for this value of $x_f = (r/R)_f$ the values for t_f and p_f and their gradients must be the same for both solutions. This requirement provides two equations from which the constants C and D can be determined (see Fig. 12.1). Schwarzschild developed an elegant way to facilitate the matching of the two solutions. His method also provides relatively good insight into the processes which

141

determine stellar structure. For simple models it works very well. The method becomes complicated when we find convection zones in stars or zones with different chemical abundances or both. Under those conditions more than two constants occur in the differential equations, and more than two different zones have to be matched. The number of trial solutions becomes large. For these reasons modern studies prefer the more flexible though less transparent calculations by means of Henyey's method. Since Schwarzschild's method is now rarely used and also because it is described very well in Schwarzschild's book, we will not describe it here in detail.

12.2   Henyey's method

12.2.1 *The basic equations*

This method was first described by Henyey *et al.* (1959).

We want to calculate the unknown functions, pressure $= P(M_r, t)$, temperature $= T(M_r, t)$, radius $= r(M_r, t)$, luminosity $= L_r(M_r, t)$, hydrogen abundance by mass $= X_1(M_r, t)$ and helium abundance by mass $= Y(M_r, t) = X_2(M_r, t)$, where t stands for time. The total mass M and the chemical abundances at time $t = t_0$ are given. Here M_r is the mass

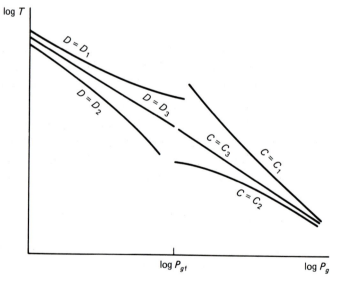

Fig. 12.1. Demonstrates the principle of Schwarzschild's method: A series of integrations is performed from the inside out assuming (basically) different constants D. Another series of integrations is performed from the outside in, assuming different values for the constant C. Only for the constants $D = D_3$ and $C = C_3$ can the solutions be fitted smoothly at the fitting point with $P_g = P_{gf}$.

inside a shell of radius r. Since the zones with given abundances are nearly stationary in mass while during stellar evolution they may move in and out in radius r, it is better to choose the mass M_r inside a given shell as the independent variable and calculate the corresponding radius, rather than use the radius as the independent variable.

In order to obtain the five unknown functions we have to solve the following differential equations (see Chapter 9) when we use M_r as the independent variable.

$$\frac{\partial r}{\partial M_r} = \frac{1}{4\pi r^2 \rho} \tag{12.1}$$

describing the change of mass M_r with an increase of radius by ∂r. This is the transformation equation.

The condition of hydrostatic equilibrium is then

$$\frac{\partial P}{\partial M_r} = -\frac{GM_r}{4\pi r^4} \tag{12.2}$$

describing the equilibrium between pressure and gravitational forces.

The energy equation is

$$\frac{\partial L_r}{\partial M_r} = \varepsilon - T\frac{\partial S}{\partial t} \tag{12.3}$$

where S is the entropy. We have $dS/T = dQ$, where Q is the internal energy. Equation (12.3) expresses the fact that the energy ε generated per g s has to either increase the luminosity, when it is transported outwards, or it has to be used to increase the thermal energy Q, or the enthalpy.

In addition we must describe the temperature stratification either for radiative energy transport or for convective energy transport. For the latter the temperature stratification is adiabatic in the stellar interior where the energy exchange is very small and the density ρ is very high (see equation (6.46)).

Generally the temperature stratification can be described by

$$\nabla = \frac{\partial \ln T}{\partial \ln P} = \frac{P}{T}\frac{\partial T}{\partial P} = \frac{P}{T}\frac{\partial T}{\partial M_r}\frac{\partial M_r}{\partial P} \tag{12.4}$$

or, using equation (12.2), we find

$$\frac{\partial T}{\partial M_r} = \nabla\frac{T}{P}\frac{\partial M_r}{\partial P} = -\nabla\frac{T}{P}\frac{GM_r}{4\pi r^4} \tag{12.5}$$

Here P is the total pressure including gas pressure P_g, radiation pressure P_r and perhaps turbulence pressure P_t.

Equation (12.5) does not contain any physics describing the kind of energy transport. The physics is introduced when calculating the actual value of ∇.

If the energy transport is radiative then

$$\nabla = \nabla_{\mathrm{rad}} = \frac{3\kappa L_r P}{16\pi\sigma T^4} \frac{1}{GM_r} \tag{12.6}$$

For convective energy transport in the interior

$$\nabla = \nabla_{\mathrm{ad}} \tag{12.7}$$

At each layer we have to check if $\nabla_{\mathrm{rad}} < \nabla_{\mathrm{ad}}$; if so equation (12.6) has to be used, otherwise equation (12.7) must be used. In low density regions we have to calculate ∇_{conv} using the convection theory for instance as described in Chapter 6.

In order to determine the abundances X, Y and Z which enter into the calculations of κ and ε, we have to give the original abundances at the time $t = t_0$ and prescribe how the abundances at a given M_r change in time due to nuclear reactions. One equation must describe how the helium abundance Y increases due to the combining of four H^1 into He^4 and the other must describe how the hydrogen abundance X diminishes because four H^1 disappear when one He^4 is formed.

If the CNO cycle contributes to the energy generation then the abundance changes of the C, N and O elements and isotopes involved in the cycle also have to be considered. If those abundances change the number of reactions changes and the energy generation changes.

If during late stages of stellar evolution heavy elements are generated these abundance changes have to be included in the ε term and possibly in the κ term.

If there is a negative μ gradient, i.e. *if the deeper layers have a larger μ than the upper layers*, we have to check carefully under which circumstances a rising bubble, which may have a larger μ due to a larger helium abundance, will still have a lower density ρ than its surroundings. *The actual temperature gradient needed for convective instability has to be larger than for constant μ*, large enough to overcome the stabilizing effect of the μ gradient. For convective instability the ∇_{rad} has to be larger than a critical gradient ∇_{cr} which is larger than ∇_{ad}, the difference depending on the μ gradient. *If μ is increasing outwards*, which might happen if helium-rich material is accreted from a neighboring star, *then $\nabla_{\mathrm{cr}} < \nabla_{\mathrm{ad}}$*. In this case we can have instability even for $\nabla < \nabla_{ad}$. This is a so-called *Rayleigh–Taylor instability*.

Equations (12.1), (12.2), (12.3) and (12.6) or (12.7) with specifications (12.5) are the four differential equations of first order which have to be solved in order to determine the four unknown functions P, T, r, L as functions of M_r. We still have to give the equation of state which relates ρ with P and T.

For the integration of the four differential equations of first order describing P, T, L and r at any given time t we need four boundary conditions. At the inner boundary coefficients containing r^{-1} become singular (see equations (12.5), (12.6)). In order to avoid this problem we apply boundary conditions at $r_j > 0$, but only slightly larger than 0. If r_j is very small then we may assume that inside r_j the ρ is constant and $\rho = \rho_c$. We then have the relations

$$r_j = \left(\frac{3}{4\pi\rho_c}\right)^{1/3} M_{r_j}^{1/3} \tag{12.8}$$

$$P_j = P_c - \frac{1}{2}\left(\frac{4\pi}{3}\right)^{1/3} G\rho_c^{4/3} M_{r_j}^{2/3} \tag{12.9}$$

Equation (12.9) describes the fact that at the radius r_j the pressure P_j is lower than the central pressure P_c by the weight of the overlying column of thickness r_j. The luminosity at this r_j must be given by the energy generated inside the radius r_j by the mass M_{r_j}, reduced by the amount of energy which might be used to increase the temperature, which means

$$L_{r_j} = \left(\varepsilon - T\frac{dS}{dt}\right)_c M_{r_j} \tag{12.10}$$

We can also describe the temperature $T(r_j)$, because it must be given by the central temperature

$$T_c - (dT/dM_r)M_{r_j} \tag{12.11}$$

which leads to

$$T_{r_j} = T_c - \frac{1}{2}\left(\frac{4\pi}{3}\right)^{1/3} G(\rho_c^{4/3}/P_c)\nabla_c T_c M_{r_j}^{2/3} \tag{12.12}$$

In this way we prescribe four boundary conditions near the center, but we do not know T_c and P_c (ρ_c is a function of P_c and T_c given by the equation of state). We therefore cannot start the integration in the center.

We also have boundary conditions at the base of the atmosphere. At this point, if the atmosphere is very thin and contains very little mass, we must have $M_r = M$, the mass of the star, which we have to give if we want to calculate a stellar model. In the atmosphere $r = R$, which has to be calculated, and $L_r = L$ which also has to be calculated. We do know, however, that

$$T_{\text{eff}}^4 = \frac{L}{4\pi R^2 \sigma} \quad \text{and} \quad T^4(\bar{\tau}) = \tfrac{3}{4} T_{\text{eff}}^4 (\bar{\tau} + q(\bar{\tau})) \qquad (12.13)$$

with $q(\bar{\tau}) \approx \tfrac{2}{3}$ (see Volume 2 or equation (7.1)). These equations relate T to L and R and therefore provide a boundary condition.

We have thus two real boundary conditions at the surface, namely $M_r = M$, and equations (12.13). In the center we have four boundary conditions with two unknowns, which also are equivalent to two boundary conditions.

At no boundary do we have enough conditions to start the integration. The Henyey method aims at making use of *all four boundary conditions simultaneously* in order to find the solution for the whole star at any given time t. The trick is to replace the differential equations by difference equations, and solve a large system of linear equations for the unknowns at all depth points.

12.2.2 The choice of variables

In order to replace the differential equations by difference equations we have to set up a number of grid points in the independent variable. We have to ensure that we have a large number of grid points in regions where P, T, κ, etc., change fast, which means close to the surface. In equations (12.1)–(12.7) we have used M_r as the independent variable, but in the outer regions close to the atmosphere P, T and ρ change very rapidly while M_r hardly changes at all. One small step in M_r would correspond to a large change in P and T. In order to keep the changes in ρ, T and P small, we would need very many grid points in M_r, which means extremely small values of ΔM_r from one grid point to the next, while in the center we do not need that many points because all the variables change slowly with M_r. An independent variable such as

$$\xi = \ln\left(1 - \frac{M_r}{(1+\eta)M}\right) \qquad (12.14)$$

gives many grid points close to the surface but relatively few points near the center. Here $\eta \ll 1$ and can be chosen arbitrarily. Equal spacings $\Delta\xi$ give many grid points in the atmosphere and fewer close to the center.

Other choices of $\xi(M_r)$ are possible of course. Eggleton (1971), for instance, chooses an independent variable which changes with time. We then have to transform to the new independent variable with each time step, but computing time can be saved in that way for a properly chosen independent variable.

We will in the following call the independent variable $\xi(M_r)$ without specifying the function. We assume that all differential equations have been transformed to differentials with respect to ξ.

For any numerical method to work accurately the dependent variables as well as the independent variables should not vary by orders of magnitude through the star. Since pressure and temperature change by many orders of magnitude it is better to choose for instance $p = \ln P$, $\theta = \ln T$, $x = \ln r$ and $l = \ln L$ as dependent variables. These new variables are then all functions of ξ.

12.2.3 Replacing differentials with differences

We now set up the grid points ξ_j, with $j = 1$ to m (see Fig. 12.2).

After all the transformations we have differential equations in all the new variables. For instance, the differential equation describing the change of radius with mass at a given layer with an index j where the grid point is ξ_j reads

$$\frac{\mathrm{d}x}{\mathrm{d}\xi} = \frac{1}{4\pi r^2(x)\rho(p,\,\theta,\,X_i)}\frac{\mathrm{d}x}{\mathrm{d}r}\bigg/\frac{\mathrm{d}\xi}{\mathrm{d}M_r} \tag{12.15}$$

Similar equations are derived from the other differential equations.

The X_i give the element abundances; the index i stands for the different elements.

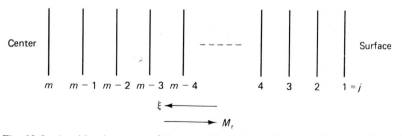

Fig. 12.2. A grid point system ξ_j is set up throughout the star. The index j runs from 1 to m. The index $j = 1$ refers to the point where the outer boundary condition is applied. m refers to the center and $m - 1$ to the point r at which the interior boundary condition is applied.

The differentials on the right-hand side can be calculated analytically from the transformation relation. They are known functions of the variables at point j. In order to do it right we have to evaluate *all* the variables at the *same* layer.

If we express the right-hand side of equation (12.15) by the variables at layer j then the derivatives at the left-hand side also have to be evaluated at the layer j. This means we have to write

$$\left.\frac{dx}{d\xi}\right|_j = \frac{x(j+1) - x(j-1)}{\xi(j+1) - \xi(j-1)} \tag{12.16}$$

which would use rather large intervals to describe the differential.

If we want to use half that spacing and write

$$\frac{dx}{d\xi} = \frac{x(j+1) - x(j)}{\xi(j+1) - \xi(j)} = \left.\frac{dx}{d\xi}\right|_{j+1/2} \tag{12.17}$$

then this actually describes the differential at the center between j and $j+1$, i.e. at the layer with index $j + \frac{1}{2}$, which is not a grid point. In order to be consistent we then have to express the right-hand side also by the variables at the point $j + \frac{1}{2}$. Since this is not a grid point for which the variables will be calculated we have to describe the variables at the point $j + \frac{1}{2}$ by means of the variables at the grid points as for instance

$$\rho(j + \tfrac{1}{2}) = \frac{\rho(j+1) + \rho(j)}{2} \tag{12.18}$$

and similarly for all the variables. If a function varies rapidly over the interval j and $j + 1$, as may be the case for ε, then other averages, for instance, the geometric mean, may better describe the value at the point $j + \frac{1}{2}$. At any given time we have for each layer with index $j + \frac{1}{2}$ a set of equations in which only the variables at the points j and $j + 1$ occur. This means at each half point we have a set of four equations, replacing the four differential equations, which are of the form

$$G_l(p_j, \theta_j, x_j, l_j, \xi_j, p_{j+1}, \theta_{j+1}, x_{j+1}, l_{j+1}, \xi_{j+1}) = 0 \tag{12.19}$$

where $l = 1, 2, 3, 4$, for the four differential equations. The G_l stand for the functions derived from the differential equations describing $p(\xi)$, $\theta(\xi)$, $x(\xi)$ and $l(\xi)$. The $j = 1, \ldots, m - 2$ if $m - 1$ is the number of grid points. (For the region m (center) to $m - 1$ we have made the analytical integration with $\rho = $ const., $\varepsilon = $ const.). This is a system of $4 \times (m - 2)$

equations for the $4 \times (m - 1)$ unknowns at the layers with $j = 1, \ldots, (m - 1)$. We need four additional equations to determine the unknowns. The additional four equations are the boundary conditions. Of course we also have to know the chemical abundances at each point. For main sequence stars we assume for the time $t = t_0$ homogeneous abundances through the whole star. Following the contraction history we see no reason why there should be any element separation. (The time scales are too short for diffusion to work through major parts of the star.)

The outer boundary condition applies to an atmospheric layer, for instance a layer with $\bar{\tau} = \frac{2}{3}$, for which $T = T_{\text{eff}}$ or $\sigma T^4 = L/4\pi R^2$. $\xi(1)$ must then refer to this layer and $M_r(1) = M$. Integrations of the hydrostatic equations for different R and L from $\tau = 0$ down to $\bar{\tau} = \frac{2}{3}$ determine how P depends on L and R for a given M. This relation provides another boundary condition. Both boundary conditions relate the variables R, T and L or P, R and L at the point $j = 1$. They only contain variables at this layer. They are of the general form

$$B_1(p_1, \theta_1, x_1, l_1) = 0 \quad \text{and} \quad B_2(p_1, \theta_1, x_1, l_1) = 0 \qquad (12.20)$$

This yields only two additional equations. We need two more, which come from the interior boundary conditions. The analytical integration for the central region from $j = m$ (center) to $j = m - 1$, the innermost shell point, leads to equations (12.8)–(12.12) which give us four additional equations. At first sight it might appear that this is more than we can use. We realize, however, that these equations contain the additional unknowns ρ_c, T_c, P_c. Making use of the equation of state, we are left with two unknowns, θ_c, p_c, and four boundary conditions at the grid point $j = m - 1$ which contain all the variables at the layer $m - 1$ and the two variables for the center p_c and T_c. The boundary conditions are of the form $C_i(p_{m-1}, \theta_{m-1}, X_{m-1}, l_{m-1}, p_m, \theta_m) = 0$, for $i = 1, \ldots, 4$. The total number of equations then equals the number of unknowns.

The equations are algebraic but non-linear. Before we can solve this system we have to linearize the equations. For this we need an approximate zero order solution $p_0(\xi)$, $\theta_0(\xi)$, $X_0(\xi)$, $l_0(\xi)$. We then write a better approximation to $p(\xi)$ as

$$p_1(\xi) = p_0(\xi) + \Delta p(\xi) \qquad (12.21)$$

and similarly for the other unknowns.

In all the equations we replace the p, θ, x, l, by these expressions and then neglect the terms which are quadratic or of higher order in the corrections Δp, $\Delta \theta$, Δx, and Δl, leaving only terms which are linear in these corrections. The equations then read

$$G_i = G_{i0} + \frac{\partial G_i}{\partial p_j} \Delta p_j + \frac{\partial G_i}{\partial \theta_j} \Delta \theta_j + \frac{\partial G_i}{\partial x_j} \Delta x_j + \frac{\partial G_i}{\partial l_j} \Delta l_j + \frac{\partial G_i}{\partial p_{j+1}} \Delta p_{j+1}$$

$$+ \frac{\partial G_i}{\partial \theta_{j+1}} \Delta \theta_{j+1} + \frac{\partial G_i}{\partial x_{j+1}} \Delta x_{j+1} + \frac{\partial G_i}{\partial l_{j+1}} \Delta l_{j+1} = 0 \qquad (12.22)$$

All the derivatives are calculated from the zero order approximation.

In addition the boundary conditions have to be linearized in the same way.

We then have a system of $4(m - 1) + 2$ equations for the $4(m - 1) + 2$ unknowns, namely the corrections Δp, $\Delta \theta$, Δx and Δl as well as Δp_c and $\Delta \theta_c$. These equations can be solved for the corrections. We thus find a new, better approximation for the unknowns at the grid points. These are not yet the correct solutions because we have used the zero order approximations p_0, θ_0, x_0 and l_0 to calculate the coefficients in the system of equations. Setting now

$$p_2(\xi) = p_1(\xi) + \Delta p(\xi) \qquad (12.23)$$

and similarly for the other unknowns we can calculate new coefficients using the better approximations p_1, θ_1, x_1, and l_1 and solve again for the Δp, $\Delta \theta$, Δx and Δl as well as Δp_c and $\Delta \theta_c$, giving us a yet better approximation to the correct solution.

If the zero order approximation was reasonably close the Δp, etc. should become small after a few steps of iteration and the solutions p_n, etc., get closer with each step and finally converge to a final solution $p(\xi)$, $\theta(\xi)$, $x(\xi)$ and $l(\xi)$, as well as p_c and θ_c.

12.2.4 Solution of the system of equations

With four unknowns at each of the $m - 1$ layers in the star we have a large system of equations with a large number of unknowns. Even for a system of linear equations it is not a trivial problem to calculate the solutions but there are computer programs available now. It is to our advantage that at each layer with index $j + \frac{1}{2}$ the equations contain only the unknowns in two adjacent layers with indices j and $j + 1$ and not the unknowns from all layers; in other words most of the coefficients in the system of linear equations are zero except the diagonal ones in the matrix of coefficients. The scheme of coefficients is shown in Fig. 12.3, which is taken from Kippenhahn, Weigert & Hofmeister (1967). Of special import-ance is the fact that in the top boundary conditions only unknowns of one point occur. This special scheme of coefficients enables us to develop a method for the solution of the system of linear equations, which is saving time and computer memory. It was first developed by Henyey *et al.* (1959), and was extremely important at the time. It is described in detail in the article by Kippenhahn, Weigert and Hofmeister (1967).

12.3   Stellar evolution

12.3.1 *Reason for stellar evolution*

So far we have considered the time t_0 to be a given value. No time changes have been considered. However, because of nuclear burning the number of particles in the interior does change with time – it decreases. This tends to reduce the pressure and therefore provokes a slight shrinking

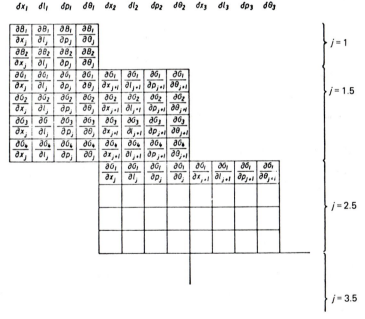

Fig. 12.3. The structure of the six equations for the top two layers in the system of linear equations to be solved. In the top layer there are two equations (the boundary conditions) for the four unknowns in layer $j = 1$. For the layer $j = \frac{3}{2}$ we have four equations for the eight unknowns in layers $j = 1$ and $j = 2$ only. For the layer $j = 2\frac{1}{2}$ we again have four equations with eight unknowns in layers $j = 2$ and $j = 3$ only. From Kippenhahn, Weigert and Hofmeister (1967).

of the star with a corresponding need to increase the central pressure and temperature. The nuclear processes force a slow change in the chemical abundances and thus in the stellar structure. For a model at a later time we have to consider a slightly different chemical composition with marginally different energy generations ε and different absorption coefficients κ. Since the star changes its density structure we must also consider the release of gravitational energy.

12.3.2 Changes in chemical abundances

If the abundances by mass of the different chemical elements i at time t_0 are called X_i and the abundances at the time $t_1 = t_0 + \Delta t$ are called X_i^*, we must have

$$X_i^* = X_i + \frac{\partial X_i}{\partial t} \Delta t$$

The change of the abundances X_i of the different chemical elements i with time are given by

$$\frac{\mathrm{d}X_i}{\mathrm{d}t} = -\sum_k \frac{\varepsilon_{ik}}{Q_{ik}} + \sum_n \frac{\varepsilon_{ni}}{Q_{ni}} \tag{12.24}$$

where the first term on the right-hand side describes the abundance change of element X_i due to nuclear reactions destroying this element. Here the ε_{ik} is the amount of energy generated per gram material and per second, and Q_{ik} is the energy gained if one gram of element i is destroyed. The ratio gives the abundance change per second. The second term on the right-hand side describes the abundance gain of element i due to reactions which lead to the formation of element i. Again ε_{ni} is the amount of energy gained per g s by this process and Q_{ni} the amount which would be gained if one gram of element i is formed. For instance if we consider the abundance change of N^{14} in the CNO cycle, the first term would describe the destruction of N^{14} by the reaction

$$N^{14} + H^1 \rightarrow O^{15} + \gamma$$

and the second term would describe the formation of N^{14} by the reaction $C^{13} + H^1 \rightarrow N^{14} + \gamma$. Q_{ni} is the energy which would be gained if one gram of N^{14} is formed, while ε_{ni} is the amount of energy actually gained per gram of material and per second, which is of course much smaller.

If the energy generation takes place in a convective region we have to take into account the fact that the material is mixed essentially instantaneously. We then have to average the abundance changes over the whole convection zone, for instance the cores of hot stars. We find

$$\frac{\partial X_i}{\partial t} = \frac{\int_{conv} \left(-\sum_k \frac{\varepsilon_{ik}}{Q_{ni}} + \sum_n \frac{\varepsilon_{ni}}{Q_{ni}} \right) dM_r}{\int_{conv} dM_r} \tag{12.25}$$

12.3.3 Gravitational energy release

If during evolution the density stratification of the star changes we have to consider the gravitational energy release, which locally is described by $-P\,dV$, where $V = \rho^{-1}$ is the specific volume. This energy will be used to increase either the radiative flux or the temperature. Considering nuclear energy generation also, we can write (see equation (10.33))

$$\frac{dL_r}{dr} = 4\pi r^2 \rho \left[\varepsilon_n + \frac{P}{\rho^2} \frac{d\rho}{dt} - \frac{d}{dt} \left(\frac{3}{2} \frac{R_g}{\mu} T \right) \right] \tag{12.26}$$

if the change in thermal energy is determined only by the change in kinetic energy. In the outer convection zone the change in ionization energy also has to be considered.

The last two terms in the bracket of equation (12.26) can be combined to give

$$\frac{dL_r}{dr} = 4\pi r^2 \rho \left[\varepsilon_n - \frac{3}{2} \rho^{2/3} \frac{d}{dt} \frac{P}{\rho^{5/3}} \right] \tag{12.27}$$

The changes over time in volume and in internal heat appear as a 'correction' term to the nuclear energy generation ε_n. In the outer stellar regions this term may be the only energy generation or consumption term. If changes over time, i.e. evolutionary changes, are calculated this term has to be taken into account in the energy equation.

12.3.4 Evolution computations

For evolution calculations we have to take time steps of small but finite length Δt. We assume a certain Δt and calculate

$$\Delta X_i = \left[-\sum_k \frac{\varepsilon_{ik}}{Q_{ik}} + \sum_n \frac{\varepsilon_{ni}}{Q_{ni}} \right] \Delta t \qquad (12.28)$$

For the new abundances we calculate a static model at time $t_1 = t_0 + \Delta t$. Since the abundance changes are so slow the star can always adjust to a new hydrostatic configuration in a much shorter time (the Kelvin–Helmholtz time) than the time over which the abundances change measurably. Once we have an approximation for the new configurations $t = t_1$, we can calculate $\Delta(P\rho^{-5/3})/\Delta t$ for every depth layer and correct the energy equation. While originally we had to take the ε_{ik} and ε_{ni} as calculated for $t = t_0$ we can now use an average value for times t_0 and t_1. We can again iterate until we find convergence. Then the next time step can be taken.

This outlines the general procedure. For details and necessary precautions concerning numerical methods and choosing the right time steps the reader is referred to the article by Kippenhahn, Weigert and Hofmeister (1967).

13

Models for main sequence stars

13.1 Solar models

Here we describe only a few representative main sequence stellar models. One is for the zero age sun, that is, the sun as it was when it had just reached the main sequence and started to burn hydrogen. We also reproduce a model of the present sun, a star with spectral type G2 V, i.e. $B - V \sim 0.63$ and $T_{\text{eff}} \sim 5800$ K, after it has burned hydrogen for about 4.5×10^9 years. In the next section we discuss the internal structures of a B0 star with $T_{\text{eff}} \sim 30\,000$ K and an A0 type main sequence star with $T_{\text{eff}} \sim 10\,800$ K. There are several basic differences between these stars. For the sun the nuclear energy production is due to the proton–proton chain, which approximately depends only on the fourth power of temperature and is therefore not strongly concentrated towards the center. We do not have a convective core in the sun, but we do have an outer hydrogen convection zone in the region where hydrogen and helium are partially ionized. The opacity in the central regions of the sun is due mainly to bound-free and free-free transitions, though at the base of the outer convection zone many strong lines of the heavy elements like C, N, O and Fe also increase the opacity.

In Table 13.1 we reproduce the temperature and pressure stratifications of the zero age sun. In Table 13.2 we give the values for the present sun as given by Bahcall and Ulrich (1987). The central temperature of the sun was around 13 million degrees when it first arrived on the main sequence; Bahcall and Ulrich calculate that the Sun has increased its central temperature by approximately 2 million degrees since. Why? During hydrogen burning four protons are combined to make one He^4. This means that after 50 per cent of the hydrogen has been transformed to helium the number of particles has decreased by a factor of 0.73, if the helium abundance was originally 10 per cent by number. For a given temperature this decreases the gas pressure by the same factor if the star did not contract.

155

Table 13.1. *Distribution of mass, temperature, pressure, density and luminosity for the young sun at the age of 5.4 × 10^7 years when it had R = 6.14 × 10^{10} cm = R_{⊙Z}, L = 2.66 × 10^{33} erg s^{-1} and T_{eff} = 5610 K. (These data were provided by C. Proffitt.)*

$r/R_{⊙Z}$	$M_r/M_⊙$	T [K]	P_g [dyn cm^{-2}]	ρ [g cm^{-3}]	$L/L_⊙$	$r/R_⊙$
0	0	13.62 (6)	1.49 (17)	8.02 (1)	0	0
0.014	1.00 (−4)	13.62 (6)	1.48 (17)	8.01 (1)	0.001	0.012
0.018	2.22 (−4)	13.60 (6)	1.48 (17)	7.99 (1)	0.003	0.016
0.035	1.64 (−3)	13.49 (6)	1.45 (17)	7.89 (1)	0.020	0.031
0.057	7.23 (−3)	13.23 (6)	1.38 (17)	7.67 (1)	0.076	0.051
0.081	1.99 (−2)	12.84 (6)	1.28 (17)	7.33 (1)	0.164	0.072
0.098	3.42 (−2)	12.49 (6)	1.19 (17)	7.03 (1)	0.233	0.087
0.115	5.32 (−2)	12.09 (6)	1.10 (17)	6.69 (1)	0.309	0.101
0.125	6.71 (−2)	11.84 (6)	1.04 (17)	6.45 (1)	0.358	0.110
0.138	8.75 (−2)	11.50 (6)	9.59 (16)	6.14 (1)	0.418	0.122
0.147	1.05 (−1)	11.24 (6)	9.00 (16)	5.90 (1)	0.461	0.130
0.158	1.26 (−1)	10.94 (6)	8.33 (16)	5.61 (1)	0.506	0.140
0.178	1.69 (−1)	10.40 (6)	7.16 (16)	5.07 (1)	0.575	0.157
0.198	2.18 (−1)	9.85 (6)	6.03 (16)	4.51 (1)	0.625	0.174
0.219	2.75 (−1)	9.28 (6)	4.93 (16)	3.92 (1)	0.655	0.193
0.263	3.99 (−1)	8.18 (6)	3.10 (16)	2.80 (1)	0.682	0.232
0.424	7.63 (−1)	5.26 (6)	4.11 (15)	5.81 (0)	0.692	0.374
0.635	9.45 (−1)	3.13 (6)	2.94 (14)	7.01 (−1)	0.690	0.560
0.731	9.74 (−1)	2.33 (6)	9.15 (13)	2.94 (−1)	0.690	0.645
0.745	9.78 (−1)	2.16 (6)	7.56 (13)	2.62 (−1)	0.690	0.658
0.843	9.93 (−1)	1.18 (6)	1.65 (13)	1.05 (−1)	0.690	0.744
1.00	1.00	5.61 (3)			0.690	0.884

The numbers in brackets give the powers of 10.

Gravitational forces therefore exceed pressure forces. The star contracts and becomes hotter. Slightly more energy is generated and the star becomes somewhat brighter. Since the arrival on the main sequence the sun has become brighter by about 0.3 magnitude. On time scales of the order of 10^9 years the sun increases its luminosity by a few per cent, while it remains on the main sequence and burns hydrogen. Stars of a given mass but different ages populate a main sequence which has a width of about 0.5 magnitude.

In Fig. 13.1 we compare the temperature and pressure stratifications for the zero age sun with the present sun. The pressure increases steeply in the center. We also show the mass distribution $M_r(r)$ in the sun. Fifty per cent of the mass is concentrated within a radius of 0.25R. In the layers in and

Table 13.2. *Distribution of mass, temperature, pressure, density, luminosity, and abundances of H, He, C and N in the present sun according to Bahcall and Ulrich (1988).*

$r/R_\odot$	$M_r/M_\odot$	T [K]	P [dyn cm^{-2}]	ρ [g cm^{-3}]	$L/L_\odot$	H	He	C	N
0.0	0.0	1.56 (7)	2.29 (17)	1.48 (2)	0.0	0.341	0.639	2.61 (−5)	6.34 (−3)
0.024	0.0014	1.55 (7)	2.21 (17)	1.42 (2)	0.012	0.359	0.621	2.50 (−5)	6.22 (−3)
0.048	0.0108	1.49 (7)	1.99 (17)	1.26 (2)	0.085	0.408	0.571	2.24 (−5)	5.98 (−3)
0.071	0.0307	1.42 (7)	1.72 (17)	1.08 (2)	0.217	0.467	0.513	1.98 (−5)	5.84 (−3)
0.095	0.0654	1.33 (7)	1.41 (17)	8.99 (1)	0.400	0.530	0.450	1.71 (−5)	5.78 (−3)
0.115	0.1039	1.25 (7)	1.18 (17)	7.64 (1)	0.553	0.577	0.403	1.50 (−5)	5.77 (−3)
0.135	0.1500	1.17 (5)	9.60 (16)	6.45 (1)	0.688	0.615	0.364	1.68 (−5)	5.77 (−3)
0.149	0.186	1.12 (7)	8.25 (16)	5.72 (1)	0.766	0.637	0.342	1.84 (−4)	5.57 (−3)
0.162	0.222	1.07 (7)	7.11 (16)	5.10 (1)	0.826	0.654	0.325	1.09 (−3)	4.52 (−3)
0.174	0.258	1.02 (7)	6.14 (16)	4.55 (1)	0.872	0.667	0.312	2.39 (−3)	3.00 (−3)
0.188	0.300	9.74 (6)	5.16 (16)	3.99 (1)	0.912	0.679	0.301	3.42 (−3)	1.80 (−3)
0.211	0.370	9.00 (6)	3.84 (16)	3.18 (1)	0.954	0.692	0.288	4.01 (−3)	1.11 (−3)
0.235	0.440	8.32 (6)	2.81 (16)	2.51 (1)	0.978	0.699	0.280	4.12 (−3)	9.86 (−4)
0.259	0.510	7.67 (6)	2.00 (16)	1.94 (1)	0.992	0.704	0.274	4.13 (−3)	9.66 (−4)
0.318	0.655	6.39 (6)	8.69 (15)	1.01 (1)	1.000	0.708	0.271	4.14 (−3)	9.63 (−4)
0.504	0.900	3.88 (6)	6.59 (14)	1.27 (0)	1.000	0.710	0.271	4.14 (−3)	9.63 (−4)
0.752	0.985	1.82 (6)	2.98 (13)	1.22 (−1)	1.00	0.710	0.271	4.14 (−3)	9.63 (−4)
0.886	0.998	6.92 (5)	2.60 (12)	2.84 (−2)	1.00	0.710	0.271	4.14 (−3)	9.63 (−4)
0.920	0.999	4.54 (5)	8.95 (11)	1.50 (−2)	1.00	0.710	0.271	4.14 (−3)	9.63 (−4)
1.000	1.000	5.77 (3)			1.00	0.710	0.271	4.14 (−3)	9.63 (−4)

The numbers in brackets give the powers of 10.

above the hydrogen convection zone we find less than 1 per cent of the total mass.

In Fig. 13.2 we show the present distribution of element abundances in the interior of the sun as calculated assuming an originally homogeneous sun with 27 per cent helium and 71 per cent hydrogen by mass at the beginning, and then burning hydrogen to helium according to local temperatures and pressures. Two per cent of the mass is in the heavy elements. Since the proton–proton chain is not very sensitive to temperature we find a small amount of hydrogen burning in off-core regions. The mass fraction in which *some* nuclear reactions occur is fairly large for low mass stars with the proton–proton chain as the energy source. For the sun we see a small change in hydrogen abundance out to 50 per cent of the mass.

We also find a change in the carbon and nitrogen abundances in the central regions where the CNO cycle is operating very slowly. Since this has been going on for 4.5×10^9 years there has been enough time to achieve equilibrium abundances of C^{12} and N^{14}, so nitrogen is about 300 times as abundant as carbon. Nitrogen is enriched by a factor of 7 while carbon has been depleted by a factor of 200. Nearly all the carbon has been transformed into nitrogen. A small fraction of oxygen has also been transformed into nitrogen. Of course, the sum of carbon, nitrogen and oxygen remains constant.

13.2   The solar neutrino problem

Generally we do not receive any radiation from the interiors of stars because all photons are absorbed and re-emitted very frequently during their diffusion to the stellar surface. There is, however, one kind of radiation for which the absorption cross-section is extremely small; these are the neutrinos. Neutrinos can therefore freely escape from the solar interior where they are formed in several nuclear reactions occurring in the energy generating cycles or chains. If we can observe these neutrinos they provide a direct check on our assertion of nuclear energy generation in stars. This observation is, of course, very difficult just because these neutrinos hardly interact with any material. So how can they be observed?

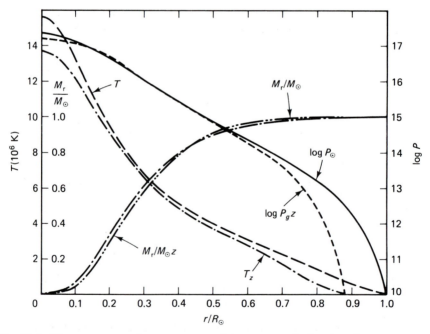

Fig. 13.1. The pressures and temperatures of the zero age (index *z*) and present sun are shown as a function of radial distance from the center. Also shown is the mass distribution in the present sun, according to Bahcall and Ulrich (1988). The data for the zero age sun were kindly provided by Charles Proffitt.

Only by using a very large amount of material and waiting for a long time and then trying to measure a very few events of interaction with the solar neutrinos.

Which kinds of neutrinos do we expect to be formed in the solar interior? In Table 13.3 we list the important reactions which occur in the different endings of the proton–proton chain. We also list the energies of the generated neutrinos because these have important implications for the possibility of observing them. Most of the neutrinos generated have very low energies, and are even more difficult to observe than the more energetic ones. There are, however, four reactions which, while inefficient for energy generation, do generate fairly energetic neutrinos. The neutrino energies from these reactions are printed in **bold** in Table 13.3.

As indicated by the long lifetimes given in the last column of Table 13.3, the reactions in lines 2 and 3 are rare in comparison with the reactions shown in line 1. The $He^3 + He^3$ reaction is the most probable ending of the proton–proton chain (see section 8.4). Only in about 14 per cent of the cases will He^3 react with He^4 and form a Be^7 nucleus. This will then lead to

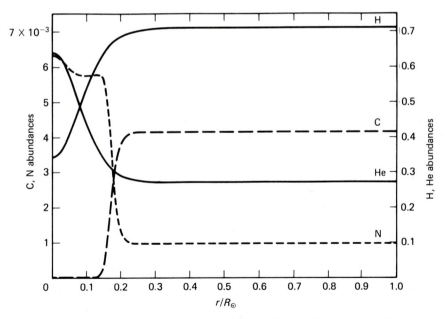

Fig. 13.2. The distributions of hydrogen, helium, carbon and nitrogen are shown as calculated for the present sun. In the center a large fraction of hydrogen has already been converted to helium. Carbon has been mainly converted to nitrogen. In the outer 50 per cent of mass, the abundances are unchanged. According to Bahcall and Ulrich (1988).

Table 13.3. *Neutrino generating nuclear reactions in the sun (from Bahcall and Ulrich (1988).)*

	Neutrino energy [MeV]	Lifetime [years]
$H^1 + p \rightarrow H^2 + e^+ + \nu_e$	0.265	10^{10}
$H^1 + p + e^- \rightarrow H^2 + \nu_e$	**1.442**	10^{12}
$He^3 + p \rightarrow He^4 + e^+ + \nu_e$	**9.625**	10^{12}
$Be^7 + e^- \rightarrow Li^7 + \nu_e$	**0.862**	10^{-1}
	0.384	
$B^8 \rightarrow Be^8 + e^+ + \nu_e$	**6.71**	10^{-8}

the generation of a neutrino with either an energy of 0.862 or 0.384 MeV or an even more energetic one with an average energy of 6.71 MeV if a B^8 nucleus is formed. The reaction $Be^7 + H^1 \rightarrow B^8 + \gamma$, however, is a factor of 1000 less probable than the reaction in line 4. We can easily estimate from the lifetimes given that these latter processes nevertheless generate most of the *energetic* neutrinos.

As pointed out above any observation of neutrinos is extremely difficult. If the neutrinos can pass entirely through the sun without interacting with any of the particles (there are about 10^{35} particles in a column of cross-section 1 cm^2 and length $R_\odot$ in the sun) then they have no problems passing through the Earth without any interaction. (If the average density of the Earth is 6 g cm^{-3} there are approximately 10^{23} particles per cm^3, and with a diameter of 12 000 km $= 1.2 \times 10^9$ cm the column density for the Earth is roughly 10^{32} particles cm^{-2}.)

Our only chance is to use particles whose cross-section for neutrino interaction is relatively large and to amass a very large amount of them, which means they must also not be very expensive.

R. Davis found that Cl^{37} has a fairly large cross-section and can be obtained inexpensively in the form of C_2Cl_4, perchlorethylene, a cleaning fluid. He built a large tank originally holding 10^5 gallons (1 gallon $\sim$ 4 litres) of this cleaning fluid. In order to cut down on interference by cosmic ray radiation, which could perturb the measurements, he placed the tank in an abandoned gold mine in South Dakota.

A neutrino can react with Cl^{37} to form A^{37} by the reaction

$$\nu_{solar} + Cl^{37} \rightarrow e^- + A^{37} \tag{13.1}$$

but only if the neutrino has an energy larger than the threshold energy of 0.814 MeV. This reaction can therefore only occur with the more energetic neutrinos as printed in bold in Table 13.3.

The most energetic neutrinos originate from the B^8 reactions. These neutrinos have the largest chance of reacting with Cl^{37} in the tank, but only 14 per cent of the He^3 reactions go through this ending. Even for these energetic neutrinos the cross-section for capture by a Cl^{37} atom is still extremely small, namely 10^{-42} cm^2. With the calculated solar flux of energetic neutrinos Davis expected to measure about six reactions per day in the 10^5 gallon tank. This means that after waiting for a week he expected to have 42 atoms of A^{37} in his 10^5 gallon tank. How long could he wait in order to accumulate more A^{37} atoms? Not longer than about a month because the A^{37} is unstable and radioactively decays back to Cl^{37} by the reaction

$$A^{37} + e^- \rightarrow Cl^{37}$$

with a half-life of 35 days. In this process the A^{37} emits an Auger electron with an energy of 2.8 MeV. So Davis had to find a way to discover about a hundred A^{37} atoms in the 10^5 gallon tank! He pumped helium gas through the tank which flushed out the A^{37}. The helium gas was then pumped through charcoal which adsorbs the A^{37}. How efficient was this method in recovering the A^{37}? In order to test this Davis added known amounts of A^{36} and A^{37} to his tank and followed the same procedure. He recovered 95 per cent of the A atoms he had added.

Davis still had to count the A^{37} atoms which were adsorbed to the charcoal. For this it was helpful that A^{37} decays by emitting an Auger electron of a well-known energy of 2.8 MeV. The charcoal was placed into an electron counting device which also measured the energy of the electron. Only those with an energy of 2.8 MeV were counted. Special precautions were taken not to measure electrons passing through the whole apparatus but only those which originated within the measuring device.

Here we have to pause and talk about the units in which neutrino fluxes are measured. The unit is SNU, which stands for *solar neutrino unit*. 1 SNU corresponds to 10^{-36} neutrino captures per target atom per second. This is equal to 1 capture per day in the 10^5 gallon tank.

With the calculated solar neutrino flux Davis expected originally to measure at least 6 SNU. The exact value depends on the adopted solar model and on the nuclear reaction cross-sections for the different reactions leading to the emission of energetic neutrinos. The most modern value for the expected number of captures is 7.9 SNU. **However, Davis only measured about 2 SNU on average after measuring for about 20 years with increasing amounts of C_2Cl_4.** This is the solar neutrino problem!

Several suggestions have been made in order to resolve the problem:

1. It has lead to speculations that the neutrino rest mass may not be zero. In this case the solar neutrinos could change into other kinds of neutrinos before they reach the Earth. This question is still unresolved.

2. Another possibility might be erroneous nuclear reaction rates. If for instance the CNO cycle is more efficient than we think or if there is an additional energy generating chain or cycle then there would be fewer proton–proton chain reactions leading to fewer energetic neutrinos. If the proton–proton chain is also more efficient than we think then the necessary energy generation could be achieved with a lower central temperature. The B^8 nuclear reactions, generating the energetic neutrinos, are very temperature sensitive ($\propto T^{18}$); a lower temperature in the solar interior could also solve the problem. If the other endings of the proton–proton cycle not including B^8 are more efficient than calculated, the solar neutrino flux is also reduced. Extensive checks of the cross-sections have lead to revisions but not to an agreement between the observed and calculated neutrino fluxes.

3. The solar model might be incorrect. This latter possibility is of special interest to us in the context of this book. As mentioned in the preceding paragraph a reduction of the solar central temperature from 15.6×10^6 K to 15.2×10^6 K could solve the neutrino problem because the generation of energetic neutrinos is so temperature sensitive. This does not sound like a large correction but it would cause several large problems. If the central temperature is decreased by 4 per cent the solar energy generation, which is proportional to T^4, is reduced by 16 per cent. The solar luminosity would then come out much too low. There is a way out. Since the sun has been burning hydrogen for 4.5×10^9 years already, hydrogen in the center is depleted by a factor up to 2 as seen in Fig. 13.2. If more hydrogen is brought into the central regions then more nuclear reactions will take place and the luminosity will increase. Therefore interior mixing could solve the luminosity problem. There is, however, also a problem satisfying the radiative transfer equation. Starting from the layer with $\bar{\tau} = \frac{2}{3}$ where $T_{\text{eff}} = 5800$ K we integrate the temperature stratification inwards and reach $T = 1.56 \times 10^7$ K in the center. In order to reach the lower temperature of 1.52×10^7 K the temperature gradient would have to be smaller. This might be achieved by choosing a larger mixing length l in the convection theory. More efficient convection would reduce

the temperature gradient and thereby the central temperature. But more efficient convection would also extend the convection zone to greater depths which gets us into trouble with the observed Li^7 abundance, which should then be zero.

The only other way to decrease the temperature gradient inwards is to reduce the radiative gradient by decreasing κ. Just below the convection zone, κ is still rather large because of the bound-free and free-free absorption of hydrogen and helium; κ is further increased by line absorption from carbon, nitrogen and oxygen. The temperature gradient and κ would be reduced were the abundance of heavy elements in these layers lower than observed at the solar surface. Ulrich (1974) found that a reduction by roughly a factor of 10 could solve the neutrino problem. A general reduction of the heavy element abundances in the solar interior seems rather unlikely. It might be easier to deplete the strongest absorbers just in the layers where they contribute most to κ; for instance, the C, N, O elements or iron in the layers below the convection zone. If, for instance, carbon would be *the* main absorber in this layer then the total radiative acceleration due to photon absorption, which transmits a momentum $h\nu/c$ to the absorbers, would work on the carbon ions alone. They would be pushed upwards into the convection zone. Carbon would then be reduced in the layer where it contributes most to κ but enriched at the surface. Oxygen might be more important in another layer and thus pushed out of that layer. In this way the main absorbers are always reduced in the layer where they perform the absorption. The κ would be reduced in all layers where elements other than helium and hydrogen are the main absorbers, and the temperature gradient would be reduced leading to a lower central temperature. If such an element, for instance carbon, is pushed up into the convection zone it is transported to the surface immediately, enriching the surface abundance. Trial calculations by Nelson and Böhm-Vitense found that if carbon were the main absorber below the convection zone the surface abundance of carbon could be increased by a factor of 2 during a time of 4.5×10^9 years while the abundance just below the convection zone is decreased by a similar factor. The assumption that the observed surface abundance holds for the deeper layers could then lead to an overestimate of κ by a factor of 4 in the important layers. While this effect must be present in principle, the quantitative results depend strongly on the actual values of the absorption coefficients for the different elements. It is not known at present whether this kind of radiative diffusion together with additional mixing in the solar interior bringing more hydrogen into the nuclear burning regions, could help to solve the neutrino problem.

Table 13.4. *Temperature, pressure and mass distribution of a zero age main sequence B0 star with M = 15 M$_\odot$, T$_{eff}$ = 30 423 K, R = 5.6 R$_\odot$.*

$r/R_\odot$	$M_r/M_\odot$	T [K]	P_g [dyn cm^{-2}]	ρ [g cm^{-3}]	$L/L_\odot$
0.0	0.014	3.04 (7)	2.06 (16)	4.50	0
0.16	0.014	3.03 (7)	2.03 (16)	4.50	4.84 (2)
0.20	0.025	3.02 (7)	2.01 (16)	4.43	8.07 (2)
0.25	0.049	3.00 (7)	1.98 (16)	4.39	1.50 (3)
0.30	0.087	2.98 (7)	1.94 (16)	4.33	2.49 (3)
0.35	0.135	2.96 (7)	1.90 (16)	4.28	3.62 (3)
0.40	0.200	2.94 (7)	1.85 (16)	4.21	4.97 (3)
0.50	0.378	2.88 (7)	1.75 (16)	4.05	7.93 (3)
0.60	0.631	2.81 (7)	1.62 (16)	3.87	1.09 (4)
0.70	0.960	2.74 (7)	1.49 (16)	3.68	1.34 (4)
0.80	1.390	2.65 (7)	1.35 (16)	3.44	1.55 (4)
1.00	2.51	2.44 (7)	1.06 (16)	2.97	1.76 (4)
1.20	3.89	2.22 (7)	7.95 (15)	2.49	1.81 (4)
1.40	5.46	1.98 (7)	5.64 (15)	2.01	1.81 (4)
1.60	7.09	1.75 (7)	3.82 (15)	1.56	1.85 (4)
2.00	10.15	1.36 (7)	1.48 (15)	7.91 (−1)	2.04 (4)
2.50	12.72	9.78 (6)	3.76 (14)	2.93 (−1)	2.25 (4)
3.00	14.07	6.98 (6)	9.53 (13)	9.84 (−2)	2.34 (4)
3.50	14.68	4.85 (6)	2.00 (13)	2.98 (−2)	2.38 (4)
4.00	14.91	3.23 (6)	3.50 (12)	7.69 (−3)	2.39 (4)
4.50	14.98	4.19 (5)	4.25 (11)	1.53 (−3)	2.39 (4)
5.00	14.99	9.15 (5)	1.83 (10)	1.39 (−4)	2.39 (4)
5.25	15.00	4.99 (5)	1.29 (9)	1.75 (−5)	2.39 (4)
5.47	15.00	1.76 (5)	1.27 (7)	6.12 (−7)	2.39 (4)
5.60	15.00	3.04 (4)			2.39 (4)

The line space shows the boundary of the convective core.
The numbers in brackets give the powers of 10.

13.3   Hot star models

In Table 13.4 we show the data for the interior temperatures and pressures of a B0 star, a hot star with $T_{eff} \approx 30\,400$ K, for the time when the star arrives at the main sequence, i.e. for the zero age main sequence B0 star. This particular star has a mass of 15 $M_\odot$ and a radius of 5.6 $R_\odot$. The data were calculated by Wendee Brunish (to whom we are indebted for making these data available to us).

In Table 13.5 we show the depth dependence of T and P for the main sequence B0 star after it has been on the main sequence for

Table 13.5. *Distribution of mass, temperature, pressure, density, luminosity, and abundances of H, He, C and N in a B0 star with 15 M$_\odot$ and age of 8.6 $\times$ 10^6 years*

$r/R_\odot$	$M_r/M_\odot$	T [K]	P_g [dyn cm^{-2}]	ρ [g cm^{-3}]	$L/L_\odot$	H	He	C	N
0.00	0	3.74 (7)	2.51 (16)	6.51	0	0.20	0.78	2.08 (−4)	1.10 (−2)
0.12	8.0 (−3)	3.72 (7)	2.46 (16)	6.50	7.31 (2)	0.20	0.78	2.08 (−4)	1.10 (−2)
0.15	1.55 (−2)	3.71 (7)	2.44 (16)	6.39	1.34 (3)	0.20	0.78	2.08 (−4)	1.10 (−2)
0.20	3.65 (−2)	3.68 (7)	2.39 (16)	6.31	2.96 (3)	0.20	0.78	2.08 (−4)	1.10 (−2)
0.25	7.00 (−2)	3.66 (7)	2.33 (16)	6.21	5.28 (3)	0.20	0.78	2.08 (−4)	1.10 (−2)
0.30	1.19 (−1)	3.62 (7)	2.26 (16)	6.08	8.25 (3)	0.20	0.78	2.08 (−4)	1.10 (−2)
0.35	1.87 (−1)	3.58 (7)	2.18 (16)	5.94	1.17 (4)	0.20	0.78	2.08 (−4)	1.10 (−2)
0.40	2.57 (−1)	3.54 (7)	2.09 (16)	5.77	1.55 (4)	0.20	0.78	2.08 (−4)	1.10 (−2)
0.50	5.16 (−1)	3.43 (7)	1.88 (16)	5.41	2.26 (4)	0.20	0.78	2.08 (−4)	1.10 (−2)
0.60	8.49 (−1)	3.31 (7)	1.67 (16)	4.99	2.84 (4)	0.20	0.78	2.08 (−4)	1.10 (−2)
0.70	1.28	3.17 (7)	1.44 (16)	4.54	3.20 (4)	0.20	0.78	2.08 (−4)	1.10 (−2)
0.80	1.79	3.01 (7)	1.20 (16)	4.10	3.28 (4)	0.20	0.78	2.08 (−4)	1.10 (−2)
1.00	3.04	2.67 (7)	8.16 (15)	3.18	3.56 (4)	0.20	0.78	2.08 (−4)	1.10 (−2)
1.20	4.33	2.33 (7)	5.23 (15)	1.43	3.58 (4)	0.36	0.62	1.07 (−4)	1.09 (−2)
1.50	5.89	1.95 (7)	3.00 (15)	1.06	3.58 (4)	0.65	0.33	6.0 (−5)	6.7 (−3)
2.00	8.33	1.50 (7)	1.27 (15)	5.74 (−1)	3.58 (4)	0.70	0.28	1.86 (−3)	2.04 (−3)
2.50	10.53	1.17 (7)	5.10 (14)	3.00 (−1)	3.58 (4)	0.70	0.28	2.81 (−3)	9.30 (−4)
3.0	12.23	9.10 (6)	1.98 (14)	1.50 (−1)	3.58 (4)	0.70	0.28	2.82 (−3)	9.20 (−4)
4.0	14.11	5.60 (6)	2.70 (13)	3.38 (−2)	3.57 (4)	0.70	0.28	2.82 (−3)	9.20 (−4)
5.0	14.77	3.41 (6)	3.36 (12)	6.81 (−3)	3.57 (4)	0.70	0.28	2.82 (−3)	9.20 (−4)
6.0	14.96	1.90 (6)	2.95 (11)	1.07 (−3)	3.57 (4)	0.70	0.28	2.82 (−3)	9.20 (−4)
7.0	14.988	8.32 (5)	9.31 (9)	7.53 (−5)	3.57 (4)	0.70	0.28	2.82 (−3)	9.20 (−4)
7.5	15.00	4.31 (5)	5.32 (8)	7.94 (−6)	3.57 (4)	0.70	0.28	2.82 (−3)	9.20 (−4)
7.95	15.00	1.27 (5)	2.79 (6)	1.25 (−7)	3.57 (4)	0.70	0.28	2.82 (−3)	9.20 (−4)
8.14	15.00	2.79 (4)			3.57 (4)	0.70	0.28	2.82 (−3)	9.20 (−4)

The line space shows the boundary of the convective core.
The numbers in brackets give the powers of 10.

8.6 $\times$ 10^6 years. Its radius has increased to 8.14 $R_\odot$, and the T_{eff} has decreased to 27 900 K. (These data were also provided by Wendee Brunish.)

For the B0 star models discussed here mass loss has not been considered. For the O stars we observe rather strong mass loss. For stars with masses larger than about 30 $M_\odot$ the mass loss may change the stellar masses measurably and the decreasing mass has to be taken into account though the exact amount is difficult to determine. Because of the mass loss, the luminosities of the stars are somewhat decreased.

In Fig. 13.3 we compare the temperature and pressure stratifications of the zero age star with 15 $M_\odot$ and the same star after it has been on the main sequence for 8.6×10^6 years. As for the sun, central temperature and pressure increase because of the conversion of hydrogen into helium and the contraction of the core. The outer layers expand because at the surface of the helium enriched core, which is at the bottom of the hydrogen envelope, the temperature has become too high for the hydrogen envelope. The surplus expands the envelope.

The temperatures in the interior of a massive star are higher than in a low mass star like the sun, but, as may be surprising at first sight, the gas pressures are not. We remember, however, from the comparison of homologous stars that we expect $P_c \propto M^2/R^4$ and $T_c \propto M/R$. If T_c increases by a factor of 2 as compared to the sun and R by a factor of 5 then the pressure must decrease, as is calculated. The reason for this is the low density of these hot stars.

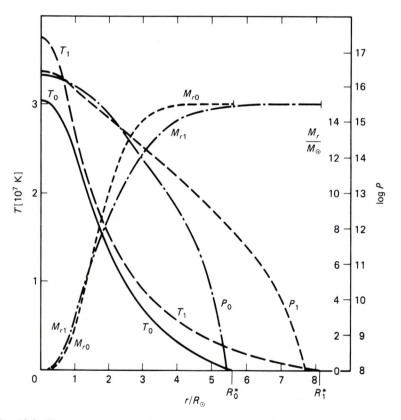

Fig. 13.3. The temperature and pressure stratifications in a zero age main sequence star with 15 $M_\odot$ are compared with the same star 8.6×10^6 years later. Due to the transformation of hydrogen into helium the central temperature, pressure and density have increased. According to data provided by W. Brunish.

We also have to be aware that for the O stars the radiation pressure P_r can be quite important. Since the total pressure $P = P_g + P_r$ has to balance the weight of the overlying material, P_g can be smaller.

Because of the higher central temperature in B stars, the nuclear energy is supplied by the CNO cycle which depends approximately on the 16th power of the temperature. The energy generation is therefore strongly concentrated towards the center. Nearly all the luminosity is generated in a very small volume. In this small volume the energy flux is very large and we find a *core convection zone*. This core therefore remains well mixed, as it has homogeneous chemical abundances in spite of the conversion of H^1 into He^4 being concentrated in the central parts of the core (see Fig. 13.4). If no convective overshoot past the boundary of the convective core occurs we should find an abrupt change in chemical abundances. In the convective core the helium abundance increases during the main sequence lifetime, while outside the core the original abundances are preserved.

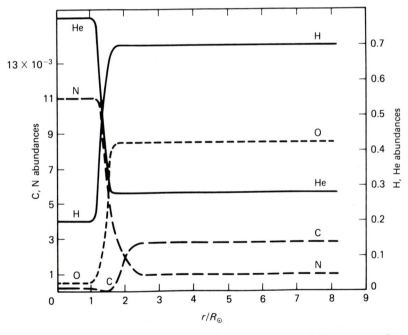

Fig. 13.4. The abundances of hydrogen, helium, carbon and nitrogen are plotted as a function of distance from the center for a star of 15 $M_\odot$ with an age of 8.6×10^6 years. In the central convection zone the abundances are uniform. The convective core extends to $r/R_\odot \approx 1.0$. Over the next half solar radius we still find abundance changes partly due to the fact that the convective core originally was somewhat larger, and partly due to semi-convection causing some mixing. Outside 2.5 $R_\odot$ the original chemical abundances are unchanged. According to data provided by W. Brunish.

Under these circumstances we may find it hard to decide for very hot stars whether the layer directly above the convective core proper becomes unstable to convection or not. We find what is called *semi-convection* (see the next section).

In the core where the CNO cycle is active the relative abundances of carbon, nitrogen and oxygen are drastically changed as seen in Fig. 13.4; we expect to see this because equilibrium abundances for the CNO cycle are established in the core.

13.4   Semi-convection

Convection sets in if the radiative temperature gradient exceeds the adiabatic one. The radiative temperature gradient increases for increasing opacity $\kappa + \sigma$, where σ is the electron scattering coefficient. In O star interiors electron scattering is most important for radiative transfer, more important than even the bound-free or free-free absorption continua. In a hydrogen atmosphere we find one electron per proton, i.e., one electron per unit atomic mass. In a helium atmosphere we find two electrons per He^4, or 0.5 electrons per unit atomic mass. The 'absorption' coefficient per gram is therefore higher in the envelope layer with 10 per cent helium than in the core with, say, 40 per cent helium. At the boundary of the well-mixed helium rich core we find by definition the boundary for convective stability using the core helium abundance; that is, we find convective stability outside of the core. In the layer just above this boundary we have, however, only 10 per cent helium. For this abundance we find $\nabla_r > \nabla_{ad}$, implying convective instability. For this low helium abundance $\kappa + \sigma$ per gram is larger and therefore ∇_{rad} is larger. So if we assume there is no mixing across the core boundary, i.e. no convection, we find that in the layers just outside of the core we have convective instability and therefore mixing. If we assume we have convective instability and therefore mixing with the core we find convective stability. So whatever we assume is inconsistent with what we find as a result of the assumption. Therefore it is generally assumed that there is some slow convection causing just enough mixing to keep this layer marginally unstable. The condition of marginal convective stability determines the helium abundance in the semi-convective layer. Even if the layer is only marginally unstable to convection we still find $\nabla = \nabla_{ad}$ for the temperature stratification. The actual degree of mixing in these semi-convective layers is still a

major point of uncertainty for the hot star stratification, as is the question how much convective overshoot we might expect. Generally the question of mixing in stellar interiors is still open to debate.

Distortion of the star due to rapid rotation, as found for most massive stars, causes so-called Eddington–Sweet circulations. Material slowly rises in polar regions and sinks in equatorial regions. Some slow mixing might occur because of this circulation; however, the increasing atomic weight towards the deeper layers with higher helium content tends to inhibit or slow down these currents so much that mixing over major fractions of the star through these currents is not theoretically expected over the lifetime of the stars.

13.5 Structure of main sequence A stars

From Volume 1 we know that early A stars with $T_{eff} \approx 10\,800$ K have masses around 2.5 $M_\odot$. Such stars are of spectral type A0. (Vega is an A0 star with $T_{eff} \sim 9500$ K.)

The A stars are different from the hot O and B stars inasmuch as they have only small convective cores. The CNO cycle is operative in the very center, but further out the temperatures are lower and the proton–proton chain still contributes to energy generation. This is therefore not as strongly concentrated towards the center as in the very hot O stars. Of course, luminosity is smaller in an A star than in an O star and this also reduces the ∇_r in the central regions and hence the size of the convective core.

The structures of A stars with $T_{eff} \sim 10\,000$ K differ from those of cooler stars because they do not have an outer convection zone which contributes any measurable convective energy transport. For all practical purposes an A star has no outer convection zone, except that small, convective velocities, of the order of cm s^{-1}, might be found, just enough to keep the outer layers well mixed.

The A stars therefore have the smallest convective regions. They have only a small convective core and essentially no outer convection zone. This is important in the understanding of magnetic A stars.

In Tables 13.6 and 13.7 we show the temperature, pressure, luminosity and chemical abundance stratifications for two main sequence A stars. (These data were kindly provided by Wendee Brunish.)

Table 13.6. *Distribution of mass, temperature, pressure, density and luminosity in a zero age main sequence star with M = 2.5 M$_\odot$, having T$_{eff}$ = 10 800 K, R = 1.97 R$_\odot$.*

$r/R_\odot$	$M_r/M_\odot$	T [K]	P_g [dyn cm^{-2}]	ρ [g cm^{-3}]	$L/L_\odot$
0	0	2.05 (7)	1.10 (17)	3.97 (1)	0
0.04	0.002	2.04 (7)	1.09 (17)	3.93 (1)	0.90
0.05	0.004	2.03 (7)	1.08 (17)	3.91 (1)	1.68
0.06	0.006	2.02 (7)	1.07 (17)	3.89 (1)	2.73
0.08	0.014	2.00 (7)	1.04 (17)	3.82 (1)	5.66
0.10	0.028	1.97 (7)	9.99 (16)	3.74 (1)	9.74
0.12	0.046	1.94 (7)	9.60 (16)	3.65 (1)	13.9
0.14	0.072	1.90 (7)	9.13 (16)	3.55 (1)	18.5
0.17	0.125	1.83 (7)	8.32 (16)	3.36 (1)	24.5
0.20	0.196	1.76 (7)	7.48 (16)	2.91 (1)	29.1
0.30	0.557	1.46 (7)	4.53 (16)	2.33 (1)	37.9
0.50	1.48	9.78 (6)	9.99 (15)	7.96	42.8
0.70	2.08	6.69 (6)	1.75 (15)	1.96	46.1
0.90	2.35	4.66 (6)	2.91 (14)	4.79 (−1)	47.6
1.10	2.45	3.26 (6)	5.11 (13)	1.19 (−1)	48.0
1.30	2.49	2.20 (6)	8.38 (12)	2.92 (−2)	48.13
1.50	2.50	1.36 (6)	1.08 (12)	6.10 (−3)	48.15
1.70	2.50	7.00 (5)	6.48 (10)	7.12 (−4)	48.15
1.91	2.50	1.45 (5)	5.36 (7)	1.93 (−5)	48.15
1.97	2.50	1.08 (4)			48.15

The numbers in brackets give the powers of 10.

13.6   The peculiar A stars

As we discussed in Volume 1 rather strong magnetic fields are observed for some early A and late B stars. Guided by solar observations we believe that the magnetic fields responsible for the *sunspot* cycle are created by a dynamo action due to the interaction of differential rotation (the equatorial regions rotate faster than the polar regions) and convection. For the magnetic late B and A stars the creation of a magnetic field by this method does not seem to be possible, at least not in the outer layers, because there is no outer convection zone with measurable convective velocities. For sunspots the suppressed convective energy transport in the magnetic regions is believed to be the origin for the lower temperatures in the magnetic spots. The formation of dark spots, similar to the sunspots, which many people claim to be present on magnetic A stars, cannot be due to the suppression of convective energy flux, because convective energy flux is not present in the outer regions of A stars and therefore cannot be

Table 13.7. *Distribution of mass, temperature, pressure, density, luminosity and abundances of H, He, C and N in a main sequence star with* $M = 2.5\,M_\odot$ *at an age of* 2.98×10^8 *years,* $R = 2.6\,R_\odot$

$r/R_\odot$	$M_r/M_\odot$	T [K]	P_g [dyn cm^{-2}]	ρ [g cm^{-3}]	$L/L_\odot$	H	He	C	N
0	0	2.50 (7)	1.34 (17)	6.37 (1)	0	0.20	0.78	1.03 (−4)	1.05 (−2)
0.03	0.0012	2.49 (7)	1.31 (17)	6.30 (1)	2.14	0.20	0.78	1.03 (−4)	1.05 (−2)
0.04	0.0028	2.47 (7)	1.30 (17)	6.25 (1)	4.70	0.20	0.78	1.03 (−4)	1.05 (−2)
0.05	0.0056	2.45 (7)	1.27 (17)	6.18 (1)	8.41	0.20	0.78	1.03 (−4)	1.05 (−2)
0.06	0.0095	2.43 (7)	1.24 (17)	6.09 (1)	1.32 (1)	0.20	0.78	1.03 (−4)	1.05 (−2)
0.08	0.021	2.38 (7)	1.17 (17)	5.89 (1)	2.46 (1)	0.20	0.78	1.03 (−4)	1.05 (−2)
0.10	0.042	2.31 (7)	1.09 (17)	5.64 (1)	3.62 (1)	0.20	0.78	1.03 (−4)	1.05 (−2)
0.12	0.070	2.23 (7)	9.94 (16)	5.35 (1)	4.55 (1)	0.20	0.78	1.03 (−4)	1.05 (−2)
0.15	0.127	2.09 (7)	8.35 (16)	4.84 (1)	5.38 (1)	0.20	0.78	1.03 (−4)	1.05 (−2)
0.20	0.265	1.81 (7)	5.87 (16)	3.33 (1)	5.72 (1)	0.36	0.62	6.18 (−5)	9.82 (−3)
0.30	0.550	1.46 (7)	3.20 (16)	1.66 (1)	5.83 (1)	0.69	0.29	1.76 (−5)	4.25 (−3)
0.50	1.24	1.01 (7)	8.78 (15)	6.53	5.87 (1)	0.70	0.28	2.18 (−3)	1.75 (−3)
0.70	1.81	7.31 (6)	2.13 (15)	2.24	5.87 (1)	0.70	0.28	2.81 (−3)	9.35 (−4)
0.90	2.15	5.43 (6)	5.20 (15)	7.37 (−1)	5.87 (1)	0.70	0.28	2.81 (−3)	9.35 (−4)
1.10	2.33	4.09 (6)	1.34 (14)	2.52 (−1)	5.87 (1)	0.70	0.28	2.81 (−3)	9.35 (−4)
1.40	2.45	2.70 (6)	1.90 (13)	5.41 (−2)	5.87 (1)	0.70	0.28	2.81 (−3)	9.35 (−4)
1.90	2.50	1.18 (6)	5.60 (11)	3.63 (−3)	5.87 (1)	0.70	0.28	2.81 (−3)	9.35 (−4)
2.25	2.50	5.17 (5)	1.51 (10)	2.23 (−4)	5.87 (1)	0.70	0.28	2.81 (−3)	9.35 (−4)
2.52	2.50	1.06 (5)	1.34 (7)	1.04 (−6)	5.87 (1)	0.70	0.28	2.81 (−3)	9.35 (−4)
2.60	2.50	9.93 (4)			5.87 (1)	0.70	0.28	2.81 (−3)	9.35 (−4)

The numbers in brackets give the powers of 10.

suppressed by magnetic fields. If dark spots are present in the peculiar A stars, there must be a different explanation.

While some magnetic fields could be generated in the convective cores of A stars and diffuse to the surface in about 10^8 years, it seems more likely that the observed magnetic fields in the magnetic A stars are just fossil fields inherited from the interstellar clouds from which the stars were formed.

14

Evolution of low mass stars

14.1 Evolution along the subgiant branch

14.1.1 Solar mass stars

From previous discussions we know that solar mass stars last about 10^{10} years on the main sequence. Lower mass stars last longer. Since the age of globular clusters seems to be around 1.2×10^{11} to 1.7×10^{11} years and the age of the universe does not seem to be much greater, we cannot expect stars with masses much smaller than that of the Sun to have evolved off the main sequence yet. We therefore restrict our discussion to stars with masses greater than about 0.8 solar masses, which we observe for globular cluster stars.

We discussed in Section 10.2 that for a homogeneous increase in μ through an entire star (due to an increase in helium abundance and complete mixing), the star would shrink, become hotter and more luminous. It would evolve to the left of the hydrogen star main sequence towards the main sequence position for stars with increasing helium abundance. In fact, we do not observe star clusters with stars along sequences consistent with such an evolution (except perhaps for the so-called blue stragglers seen in some globular clusters which are now believed to be binaries or merged binaries). Nor do we know any mechanism which would keep an entire star well mixed. We therefore expect that stars become helium rich only in their interiors, remaining hydrogen rich in their envelopes. Since nuclear fusion is most efficient in the center where the temperature is highest, hydrogen depletion proceeds fastest in the center. Hydrogen will therefore be exhausted first in the center. As helium becomes enriched, the core keeps shrinking and heating up and we find a growing helium rich core. For this small but growing helium rich star (the core) the central temperature increases; in fact its increase is sufficient to enable these already hydrogen depleted layers to generate enough energy to make up for the radiative energy loss at the surface. As these layers heat up, the CNO cycle becomes more important, relative to the proton–proton chain. Because the CNO cycle is proportional to T^{16}, the energy generation is concentrated in the highest temperature regions in which hydrogen is still present. This means that energy generation is increasingly concentrated in a narrow region near the rim of the helium rich core. A so-called shell source develops around a core which finally becomes a pure helium core containing more than 10 per cent of the stellar mass. This shell source reaches a temperature of about 20×10^6 K for which the CNO cycle is the main fusion process. Gravitational forces so squeeze the helium core that this temperature is reached in the shell source and enough energy is generated.

On the other hand we know that for a hydrogen star of one solar mass such a high interior temperature and corresponding pressure are too high to be in equilibrium with gravitational forces. While for the helium star in the center with large μ such a high temperature is needed for the balance of forces it is too high for the low μ hydrogen envelope. At some point outside the helium core the high temperature causes an excess pressure for the hydrogen envelope which then expands. Temperature, pressure and gravitational forces in the envelope are then reduced until a balance is re-established between pressure and gravity forces. Such an expansion starts already while the star is still on the main sequence but it becomes more noticeable if more than about 10 per cent of the stellar mass is in the helium core. The expansion of the hydrogen envelope terminates the main sequence lifetime, thus only 10 per cent of the hydrogen is used for energy generation during the main sequence phase of a star's lifetime.

During the phase of 'rapid' envelope expansion the shell source at the rim of the helium star does not notice anything. It maintains the high temperature necessary to replenish, by nuclear reactions, the energy lost at the surface. (Were it not to do so the thermal energy would fall below its equilibrium value, the neighboring layers would contract and heat up until enough energy is generated.) The luminosity remains essentially constant while the envelope expands. With increasing radius but constant luminosity the effective temperature must decrease. The star moves to the right in the HR diagram along the giant or subgiant branch depending on the stellar mass. During this time the shell source slowly burns its way out, getting closer to the surface and moving the 20×10^6 K layer further out in the envelope. This causes further expansion. A small increase in mass of the helium core causes a relatively large expansion of the envelope. During this phase the expansion proceeds very quickly. The stars evolve rapidly through the range with about 6500 K $> T_{eff} > 5000$ K, the so-called Hertzsprung gap, in which, consequently, very few stars are found.

As the surface temperature of the star decreases because of expansion, the hydrogen convection zone extends deeper into the star (see Chapter 7 and Volume 2). If the effective temperature of the giant decreases below about 5000 K the convection zone may finally reach down into a layer with some nuclear processed material and bring it up to the surface, where it may be observed.

14.1.2 Stars with $M \gtrsim M_{\odot}$

Stars with masses larger than 2 $M_{\odot}$ have a small convective core while on the main sequence. The convective core remains well mixed. With increasing helium abundance this convective zone shrinks, however, and leaves in its surroundings some nuclear processed material. This effect is seen in Fig. 14.1 between the mass fractions q_0 and q_1, with $q = M_r/M$.

Outside the convective core the temperature is still high enough ($T > 8 \times 10^6$ K) for the CNO cycle to proceed slowly. While it is too slow

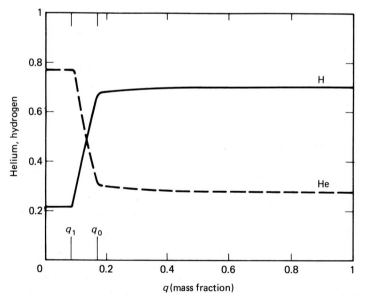

Fig. 14.1. The hydrogen and helium abundances for a star with 2.5 $M_{\odot}$ and an age of 3×10^8 years are plotted as a function of $q = M_r/M$(star). At zero age the convective core extended to $q_0 = 0.174$. At the age of 3×10^8 years it extends to $q_1 = 0.08$. The mass layers between q_1 and q_0 show hydrogen depletion and helium enrichment because of the larger extent of the convective core at the earlier times. According to data provided by W. Brunish.

to change the helium abundance, it still changes the CNO abundances (see Fig. 14.2). For the layers with $0.4 < q < 0.6$, the temperatures are too low and the nuclear reactions too slow for equilibrium abundances to be established during the age of the star. For the interior with $q < 0.4$, the temperature exceeds 10^7 K and 3×10^8 years are sufficient to establish equilibrium abundances. For $q < q_0$, the effect of the originally extended and then shrinking convective core is seen. The CNO abundances are determined by the high central temperatures which were present when the convection zone reached out to the q value under consideration. For $q < q_1$, the abundances determined by the present high central temperatures are seen.

The oxygen abundances are changed only when the CNO bi-cycle is operating, which requires high temperatures. They are therefore only influenced in the high temperature central regions and so show the effect of the shrinking extent of the convective core.

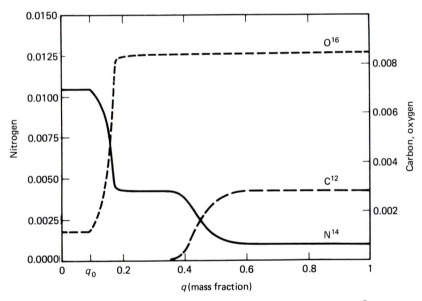

Fig. 14.2. The CNO abundances for a 2.5 $M_\odot$ star at an age of 3×10^8 years are plotted as a function of q. The nitrogen abundance is increased out to $q < 0.6$ because the temperatures inside this layer are high enough to maintain the CNO cycle, though at a very slow pace, but still slowly altering the abundances. For $q < 0.4$ the equilibrium abundances for these low temperatures are reached. Between q_0 and q_1 the steep increase due to the convection zones of various extents are seen. For $q < q_0$ the nitrogen abundance is constant for the core convection zone. The oxygen abundances are changed only for $q < q_0$. For $q > 0.6$ the nuclear reactions including oxygen are too slow to cause abundance changes at these 'low' temperatures. Data provided by W. Brunish.

By the time the surface temperature approaches 5000 K the hydrogen convection zone extends very deeply into the envelope, reaching high temperature regions. The total energy transport is then increased because convective energy transport by mass motion is more efficient than radiative energy transport by the diffusion of photons. The star therefore loses much more energy, increasing its luminosity. Energy generation is accelerated, the shell source burns upwards faster and the stellar envelope expands faster (but the radius still remains smaller than it would be for radiative equilibrium). With increasing energy transport outwards the star is still able to keep its surface temperature nearly constant. It moves up the red giant branch close to the Hayashi track. Since in the interior at the rim of the helium core the star has a temperature of about 20×10^6 K and the temperature decreases outwards with the adiabatic gradient or even less, the star cannot reach a lower temperature at the surface for the same reasons as discussed in Chapter 12 in connection with the Hayashi track.

At the same time as the envelope expands and the luminosity increases, the helium core (which is essentially a helium star in the center) grows in mass and therefore increases its central temperature which controls the stellar evolution at these phases.

During these low surface temperature, red giant phases the outer convection zones reach deep enough to dredge up material in which the CNO cycle has been operating. Enlarged ratios of N^{14}/C^{12} and C^{13}/C^{12} can therefore be expected in the atmospheres of such stars. These have been found by Lambert and Ries (1981) for cool subgiants, red giants and red supergiants, confirming that the CNO cycle is indeed operating in the interiors of these stars.

14.2   Advanced stages of low mass stellar evolution

With increasing temperature of the central helium star, new nuclear reactions may take place. In Section 8.3 we discussed the tunnel effect, necessary for nuclear reactions to occur. We pointed out that tunnelling becomes more difficult for higher Z values of particles because of larger repelling Coulomb forces, but also that tunnelling becomes easier for higher temperatures. For relatively low temperatures only reactions between particles with low Z values are possible while for higher temperatures reactions for higher values of Z also become possible (see equation (8.4)). When the helium core reaches a size of about 0.45 $M_\odot$ the envelope has greatly expanded, convection has become very efficient and the luminosity has increased so much that the star appears at the top of the red giant branch in the HR diagram. At this point the central temperature in the helium core has reached a value $T_c \sim 10^8$ K. This is high enough for the triple-alpha reaction to take place (see Section 8.6). Three He^4 nuclei combine to form C^{12}. The energy liberation per unit mass in this process is about 10 per cent of that liberated by hydrogen burning, as we inferred previously from Table 8.1.

Before we can understand what happens at this point we have to look into the equation of state for the helium core which in the mean time for low mass stars has reached such high densities that the electrons are in a so-called degenerate state.

14.3   Degeneracy

We talk about degenerate matter if, for a given temperature, the density is so high that the well-known equation of state for an ideal gas, $P_g = \rho R_g/T\mu$, breaks down. For high densities the Pauli principle (from quantum theory) becomes important for the relation between pressure, temperature and density. Pauli noticed that particles with spin $= \frac{1}{2}n$, where n is an odd number, follow different statistics than other principles. These particles are called Fermi particles. The Pauli principle states that there cannot be two or more Fermi particles with all equal quantum numbers in one quantum cell. This applies for instance to electrons, protons, neutrons but not to helium nuclei, for which the nuclear spin is 0.

A quantum cell is defined in phase space, i.e. in the six-dimensional space of geometrical space x, y, z and momentum space p_x, p_y, p_z. For a quantum cell the (volume) element in phase space is given by

$$\Delta x \Delta y \Delta z \Delta p_x \Delta p_y \Delta p_z = h^3 \tag{14.1}$$

where h is Planck's constant. The number of electrons in this quantum cell can at most be two. These two electrons must have opposite spin directions.

We now look at the geometrical volume of 1 cm^3. For a quantum cell we then have $\Delta p_x \Delta p_y \Delta p_z = h^3$, and in this momentum space there can be at most two electrons.

For the electrons in 1 cm^3 we plot in the momentum space all the arrows for a given absolute value of p between p and $p + \Delta p$ (see Fig. 14.3). All these arrows end in a spherical shell with radius p and thickness Δp. The number of quantum cells N_q corresponding to these momenta between p and $p + \Delta p$ is then given by

$$N_q = 4\pi p^2 \Delta p/h^3 \tag{14.2}$$

and the number of electrons per cm^3 which have momenta between p and $p + \Delta p$ can then at most be

$$n_e(p, \Delta p) \leqslant 2N_q = 8\pi p^2 \Delta p / h^3 = 8\pi m_e^3 v^2 \Delta v / h^3 \qquad (14.3)$$

Here m_e = electron mass. With $p = m_e v$ this corresponds to a velocity distribution. It has to be compared with the Maxwell velocity distribution which also gives the number of electrons per cm^3 with velocities between v and $v + \Delta v$, namely

$$n_e(v, \Delta v) = n_e \frac{4 \exp\left(-\dfrac{m_e v^2}{2kT}\right)}{\sqrt{\pi}\left(\dfrac{2kT}{m_e}\right)^{3/2}} v^2 \Delta v \qquad (14.4)$$

which we are accustomed to use for low densities. This is, however, correct only as long as the number of electrons is small enough that condition (14.3) is not violated. For very high electron densities the Maxwell distribution may give us a larger number $n_e(v, \Delta V)$ than is permitted by the Pauli principle, equation (14.3).

In Fig. 14.4 we have plotted the Maxwell energy distribution for electron density $n_e = 1.5 \times 10^{23}$ cm^{-3} and temperature $T = 10^5$ K. We have also plotted the maximum possible number of electrons per cm^3 with a given velocity p/m_e. For this electron density of 1.5×10^{23} cm^{-3} we do not find more electrons with a given velocity in 1 cm^3 than permitted by the Pauli principle. In Fig. 14.5 we have plotted the Maxwell velocity distribution for $n_e = 3 \times 10^{23}$ cm^{-3} together with the distribution of $2N_q$. For

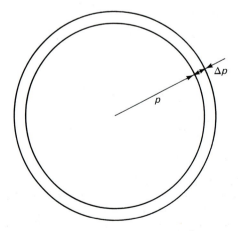

Fig. 14.3. The phase space volume for 1 cm^3 and momenta between p and $p + \Delta p$ equals the volume of the spherical shell with radius p and thickness Δp.

this density the Maxwell distribution gives us more electrons per cm³ with a given velocity than are permitted by the Pauli principle (see the cross-hatched area in Fig. 14.5). These electrons then cannot have such low velocities, and must acquire higher energies than given by the Maxwell distribution. We can only squeeze that many electrons into 1 cm³ if this additional energy is supplied. If the electrons in the cross-hatched area of Fig. 14.5 obtain the lowest possible energies they will appear in the shaded area which must be just as large as the cross-hatched area in order to accommodate all the electrons which are not permitted to be in the cross-hatched area. The actual momentum distribution obtained for this density is shown by the heavy line in Fig. 14.5. Up to a certain momentum p_0, which is called the Fermi momentum, the distribution follows the maximum possible number of electrons $2N_q$. There is also a fraction of high energy electrons which still follows the Maxwell distribution in the so-called Maxwell tail. The gas considered here would be called partially degenerate, because the Maxwell tail is still rather extended.

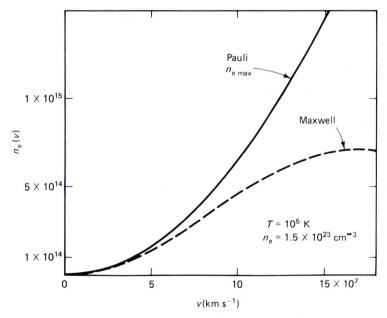

Fig. 14.4. The Maxwell velocity distribution for $T = 10^5$ K and $n_e = 1.5 \times 10^{23}$ cm⁻³ is shown. Also shown is the upper limit $n_{e\,max}$ for the number of electrons per cm³ with a given velocity permitted according to the Pauli principle (both for $\Delta v = 1$ cm s⁻¹). For $n_e = 1.5 \times 10^{23}$ the number of electrons with a given velocity v as given by the Maxwell velocity distribution never exceeds the maximum number permitted by the Pauli principle.

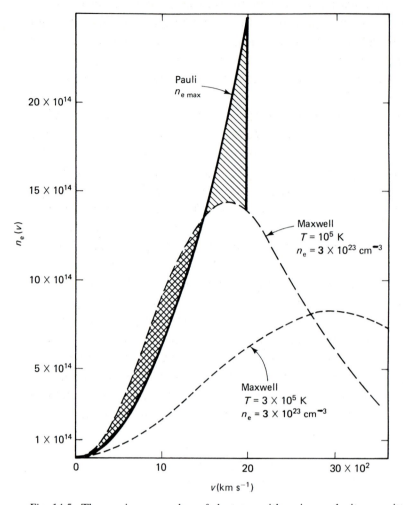

Fig. 14.5. The maximum number of electrons with a given velocity permitted by the Pauli principle is shown by the thick, solid line. Also shown is the Maxwell velocity distribution for $T = 10^5$ K as in Fig. 14.4 but now for an electron density $n_e = 3 \times 10^{23}$, i.e. twice as high as in Fig. 14.4. For this electron density the numbers given by the Maxwell distribution exceed the number $n_{e\,max}(p)$ permitted by the Pauli principle by the electron numbers included in the cross-hatched area. These electrons have to obtain higher energies and must appear in the shaded area. We also show the Maxwell velocity distribution obtained for the same $n_e = 3 \times 10^{23}$ cm^{-3} but for $T = 3 \times 10^5$ K. For $T = 3 \times 10^5$ K the Maxwell energy distribution no longer violates the Pauli principle; all the electrons have, according to the Maxwell energy distribution, higher velocities and so fewer electrons have small velocities. There is no degeneracy at this temperature. For a given electron density degeneracy can be removed by a higher temperature.

If the electron density is increased further the shaded area grows and a large number of electrons have to acquire higher energies. p_0 grows, and the Maxwell tail shrinks. *If the Maxwell tail contains a negligible number of electrons we talk about complete electron degeneracy.*

In Fig. 14.5 we have also plotted the Maxwell energy distribution for the same electron number of 3×10^{23} cm^{-3} but for a temperature of 3×10^5 K. For this higher temperature all the electrons obtain higher energies. The number of electrons with a given velocity according to the Maxwell velocity distribution never exceeds the number permitted by the Pauli principle (which does not depend on T). For this temperature the gas with $n_e = 3 \times 10^{23}$ cm^{-3} is not degenerate even though it was at $T = 10^5$ K. *For a given n_e there is always a temperature by which degeneracy can be removed.* If the temperature is high enough the number of electrons given by the Maxwell velocity distribution will stay below the number $2N_q$.

We now look at the situation for protons or neutrons. Will they also be degenerate if the electrons are degenerate? The maximum number n_n of neutrons with momenta between p and $p + \Delta p$ is again given by equation (14.3). With $p = m_n v$ and $\overline{m_n v^2} = \overline{m_e v^2}$ for equal temperature in a non-degenerate gas, we find $\bar{v} \propto 1/\sqrt{m_n}$ and $\bar{p} \propto \sqrt{m_n}$. For neutrons the momenta are all larger and therefore the number of quantum cells in momentum space is much greater for the heavy particles. According to equation (14.3) the permitted number of particles with a given velocity increases with

$$\frac{\Delta p p^2 (\text{neutron})}{\Delta p p^2 (\text{electron})} \propto \left(\frac{m_n}{m_e}\right)^3$$

(For low velocities the number of particles with a given v according to the Maxwell distribution increases only proportional to $m^{3/2}$.)

In Fig. 14.6 we have plotted the Maxwell velocity distribution for the number of *neutrons* per cm^3 equal to 3×10^{23} cm^{-3} and also the velocity distribution according to equation (14.3), taking into account the Pauli principle. For this number of particles per cm^3 we found degeneracy for electrons, while the number of neutrons stays well below the Pauli principle limit for all velocities. The velocities are lower by a factor 100 than in Figs. 14.5 and 14.4. Because of the higher masses and larger momenta we can have much higher particle densities for protons and neutrons (about a factor 10^5) before they become degenerate. Of course, in the dense helium core we do not have protons or neutrons, and helium does not obey the Pauli principle. The alpha particles do not degenerate; **only the electrons in the helium core degenerate.**

14.4   Equation of state for complete degeneracy

Since the gas pressure is determined by the kinetic energy of the particles we expect, for a given temperature, a higher gas pressure for a degenerate gas than for an ideal gas because the kinetic energies for the degenerate particles are much higher than they would be for a Maxwell energy distribution. For a completely degenerate gas we can calculate the Fermi momentum p_0, i.e. the momentum up to which all quantum cells are filled. This depends only on the electron density n_e, and not on the

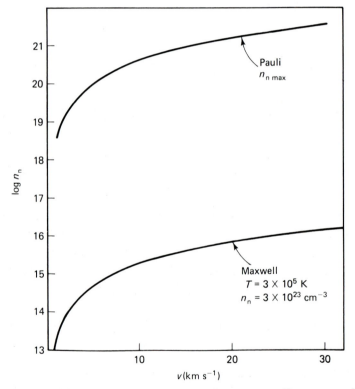

Fig. 14.6. For a particle density of neutrons $n_n = 2 \times 10^{23}$ and $T = 10^5$ K we plot the Maxwell velocity distribution. Also plotted is the maximum number of neutrons $n_{n\,max}$ with a given velocity permitted according to the Pauli principle. For a given velocity the momenta of the neutrons are larger by the ratio of the masses, therefore $n_{n\,max}(v) \gg n_{e\,max}(v)$. Notice that the $\log n_n$ are plotted because otherwise the Maxwell velocity distribution and the Pauli limit could not have been shown on one plot. Also notice that the velocities here are lower by a factor 100 than in Figs. 14.5 and 14.6.

temperature. The temperature controls only the very high energy Maxwell tail, which for complete degeneracy has a negligible number of particles. Knowing the Fermi energy corresponding to p_0 we can calculate the average kinetic energy of the particles. From this average kinetic energy the gas pressure for completely degenerate gas can be calculated as follows.

The maximum number of electrons n_e per cm^3 with momenta up to p_0 is given by the volume in phase space (see Fig. 14.3) divided by the size of a quantum cell, i.e. by h^3, multiplied by 2, which means

$$n_e(p_0) \leqslant \int_0^{p_0} \frac{4\pi p^2 \cdot 2}{h^3}\, \mathrm{d}p = \frac{8\pi}{3} \frac{p_0^3}{h^3} \tag{14.5}$$

In order to calculate the pressure for a given density we have to relate density ρ and p_0.

If $\mu_E m_H$ is the mass (of heavy particles) per electron then

$$m_H \mu_E = \frac{\rho}{n_e} \quad \text{or} \quad n_e = \frac{\rho}{\mu_E m_H} \tag{14.6}$$

(For He^{2+}, $\mu_E = 2$; for H$^+$, $\mu_E = 1$.) Making use of equations (14.5) and (14.6) we find

$$\frac{1}{\mu_E} \rho = \frac{8\pi}{3} \frac{m_H}{h^3} p_0^3 \tag{14.7}$$

In order to calculate the pressure we have to calculate the momentum transferred to a hypothetical wall in the gas if the electrons coming from all directions are reflected on the wall. The momentum transfer δp per electron with momentum p_x is given by $\delta p_x = 2p_x = 2p(p_x/p)$. The number of electrons with p and p_x arriving at the wall per cm^2 s is given by

$$n_e = \mathrm{d}n_e(p, p_x) v_x = \mathrm{d}n_e(p, p_x) \frac{p}{m_e} \cos\theta \tag{14.8}$$

where θ is the angle between p and the x-axis (see Fig. 14.7).

The $\mathrm{d}n_e(p, p_x)$ is given by the volume in momentum space occupied by the electrons with p and p and $p + \mathrm{d}p$ and p_x multiplied by 2 and divided by h^3. We find

$$\mathrm{d}n_e(p, p_x) = \frac{4\pi}{h^3} p^2\, \mathrm{d}p \sin\theta\, \mathrm{d}\theta \tag{14.9}$$

The total momentum transfer of these electrons to the wall is then

$$\delta P_{\mathrm{e}}(p, p_x) = \frac{4\pi}{h^3} p^2 \frac{p}{m_{\mathrm{e}}} \, dp \cos\theta \sin\theta \, d\theta \, 2p_x = \frac{8\pi}{m_{\mathrm{e}} h^3} p^4 \, dp \cos^2\theta \sin\theta \, d\theta$$

$$(14.10)$$

Integration over all angles θ (over the half sphere out of which particles hit the wall) gives

$$\Delta P_{\mathrm{e}}(p) = \frac{8\pi}{m_{\mathrm{e}} 3h^3} p^4 \, dp \tag{14.11}$$

In order to calculate the total electron pressure we have to integrate over all momenta p up to the Fermi momentum p_0, i.e.

$$P_{\mathrm{e}} = \frac{8\pi}{3m_{\mathrm{e}} h^3} \int_0^{p_0} p^4 \, dp = \frac{8\pi}{15 m_{\mathrm{e}} h^3} p_0^5 \tag{14.12}$$

Using equation (14.7) we find

$$p_0^5 = \left(\frac{3}{8\pi} \frac{h^3}{m_{\mathrm{H}}} \right)^{5/3} \frac{\rho^{5/3}}{\mu_{\mathrm{E}}^{5/3}} \tag{14.13}$$

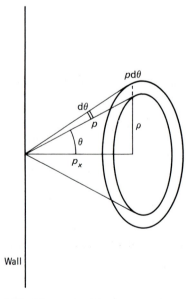

Fig. 14.7. Electrons with momentum p passing through a ring of thickness $p \, d\theta$ and radius $\rho = p \sin\theta$ hit the wall and each transfer a momentum $2p_x$ to the wall when they are reflected.

and

$$P_e = K_1 \left(\frac{\rho}{\mu_E} \right)^{5/3}$$ (14.14)

with

$$K_1 = \frac{h^2}{20 m_e m_H} \left(\frac{3}{\pi m_H} \right)^{2/3} = 9.91 \times 10^{12} \quad [\text{cgs}]$$ (14.15)

If more and more electrons are squeezed into a given volume, p_0 must increase and finally the vast majority of the electrons will have velocities very close to the velocity of light, c. We then talk about *relativistic degeneracy*. In this case $v_x = c(p_x/p) = c \cos \theta$ and the integral for the derivation of the electron pressure takes the form

$$P_e = \int_0^{P_0} \tfrac{1}{3} p c \frac{8\pi}{h^3} p^2 \, dp = \frac{8\pi}{3h^3} c \tfrac{1}{4} p_0^4 = \frac{2\pi c}{3h^3} p_0^4$$ (14.16)

or

$$P_e = K_2 \left(\frac{\rho}{\mu_E} \right)^{4/3}$$ (14.17)

where

$$K_2 = \frac{hc}{8 \mu_H} \left(\frac{3}{\pi m_H} \right)^{1/3} = 1.231 \times 10^{15} \quad [\text{cgs}]$$ (14.18)

In order to obtain the total gas pressure P_g we have to add the pressure P_H of the heavy particles, such that

$$P_g = P_e + P_H$$ (14.19)

The electron pressure P_e for a gas with degenerate electrons will, however, be much larger than P_H because the kinetic energies of the electrons have so increased because of the degeneracy, that for complete degeneracy the pressure of the heavy particles can be neglected.

A very important point is the fact that for complete degeneracy according to equations (14.14) and (14.17) the **electron pressure does not depend on the temperature, but only on the density.** Since $P_H \ll P_e$ for complete degeneracy of the electrons, the gas pressure $P_g = P_e + P_H$ is also independent of the temperature and **depends only on the density.** In Fig. 14.8 we show in the T, ρ plane roughly the regions for which electron degeneracy becomes important and where relativistic degeneracy sets in.

For the solar interior with $\rho_c \sim 10^2 \text{ g cm}^{-3}$ and $T_c \sim 1.5 \times 10^7$ K degeneracy is not yet important, although an increase in ρ by a factor of 10 would cause degeneracy.

14.5   Onset of helium burning, the helium flash

14.5.1   *Stars with solar metal abundances*

We saw that because of the diminishing number of particles due to the nuclear reactions in the core and later in the shell source the burned-out helium core contracts and increases its temperature. When stars with masses less than about 2.25 $M_\odot$ reach the tip of the red giant branch the stellar helium core has contracted so much that the electrons have become completely degenerate. This means that the pressure is due to degenerate electron pressure and is temperature independent, where the temperature is now defined by the energy distribution of the very sparsely populated high energy tail of the Maxwell energy distribution of the electrons, and **by the energy distribution of the heavy particles,** which are the ones that make nuclear reactions. **Their temperature is therefore very important,** but gravitational forces are balanced by electron pressure which does not depend on temperature. This has important consequences for the stability of the hydrostatic equilibrium. For temperatures around 10^8 K the triple-alpha reactions, combining three He^4 to one C^{12}, start in the very dense

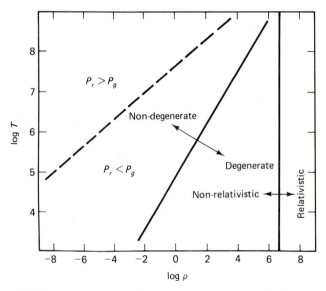

Fig. 14.8. In T, ρ plane we indicate the regions where electron degeneracy becomes important and where relativistic degeneracy is achieved. Also indicated is the region where the radiation pressure becomes important. According to Schwarzschild (1958).

core near the center. These processes generate energy and heat the core, which means they increase the kinetic energy of the heavy particles. They then make more nuclear reactions, further increasing the energy production, etc. The core heats up rapidly. If the pressure were temperature dependent the increased T would lead to an increased pressure, the core would then expand and cool off, thereby reducing the number of nuclear reactions to the equilibrium value. Because the degenerate electron pressure is independent of temperature this does not happen. The core does not expand but energy generation and heating continue to increase in a runaway situation, which is called the helium flash. During this time the interior temperature changes within seconds; the star changes faster than a computer could follow around 1960, when the helium flash was discovered.

With increasing temperature the Maxwell tail of the electron velocity distribution becomes, however, more and more populated (see Fig. 14.5), and for still higher temperatures most of the electrons again follow the Maxwell velocity distribution. The degeneracy is removed. The pressure increases again with increasing temperature, causing the core to expand and prevent further increase in temperature. At this point the star is able to find a new equilibrium configuration with an expanded non-degenerate hot helium burning core. The result is that the hydrogen burning shell source is also expanded and has a lower density and temperature and

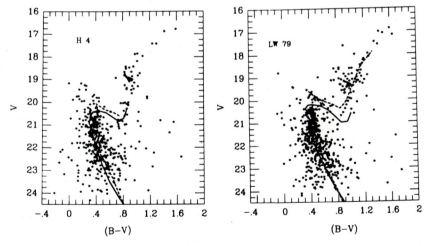

Fig. 14.9. The color magnitude diagrams for the two star clusters H4 and LW79 in the Large Magellanic Clouds. After the helium flash the low mass stars finish half-way up the red giant branch where they form the group of clump stars in the HR diagrams of globular clusters. They are burning helium in their cores. For the two clusters shown the clump stars can be recognized at $m_v = V \approx 19$ and $B - V \sim 1$. From Mateo (1987).

generates less energy from hydrogen burning, while some energy is generated in the core by the triple-alpha reaction.

Since at the bottom of the hydrogen envelope the temperature is no longer so high, the envelope shrinks and the star becomes hotter at the surface, though not that much so since at the same time its luminosity is decreasing.

What can we observe from the helium flash? Not very much, because it takes the radiation at least a thousand years to get to the surface! By that time the effect is smoothed out. We expect to see a slight increase in luminosity for a short period of time before the star decreases in luminosity. In the HR diagram the star ends up in the lower part of the giant branch where such stars form the so-called clump stars. Stars stay a relatively long time in this region while they are burning helium in their centers. That is why there are many stars at this luminosity and why they form a 'clump' in the HR diagram (see Fig. 14.9).

For stars more massive than about 3 $M_\odot$ the helium core never becomes very degenerate. Helium burning therefore starts slowly in a quasi-equilibrium configuration. These stars do not experience a helium flash.

For stars less massive than 0.5 $M_\odot$ the helium core will never become hot enough and helium burning will never start.

14.5.2 Metal poor globular clusters

In metal poor clusters we do not see a clump of stars at the giant branch of their HR diagram but instead see a horizontal branch (HB) which merges with the giant branch at about the luminosity of the clump stars (see Figs. 1.7 and 1.8). We still do not know how the stars get to the horizontal branch position in the HR diagram, but we can calculate equilibrium configurations that will put them there (see Faulkner *et al.*, 1965). These are stars with hydrogen burning shell sources and helium burning helium cores. The stars can have only thin hydrogen envelopes. The more mass is lost from the hydrogen envelope, the hotter the remaining star. Stars with the smallest amount of mass in the hydrogen envelope populate the blue part of the horizontal branch. The more mass left in the envelope, the redder the stars are. Stars with larger heavy element abundances generally appear more to the red side of the HB. We can consider the clump stars as the outermost red part of the horizontal branch. What we do not know is why and how metal poor stars lose so much mass apparently after the helium flash while metal rich ones do not seem to do so. The helium cores for all stars at the moment of the helium flash appear to have the same mass because the mass of the helium core determines the central temperature. Stars on the blue part of the horizontal branch appear to have a smaller total mass than the stars on the red end. The blue horizontal branch stars appear to have around $0.55\,M_\odot$ while those on the red giant branch of globular clusters have about $0.8\,M_\odot$, as inferred from the ages of the clusters.

14.6   Post core helium burning evolution

Since the triple-alpha reaction is even more temperature dependent ($\varepsilon \propto T^{30}$) than the CNO cycle, energy generation is even more centrally condensed, but not all energy is generated in the core. There is still the hydrogen burning shell source. The inner part of the helium star will soon consist only of carbon, and the carbon core slowly grows while a helium burning shell source also develops which burns its way out. In the meantime the hydrogen burning shell source also burns its way out and gets closer to the surface while the helium shell grows; this again causes the envelope to expand and the population I star evolves up the red giant

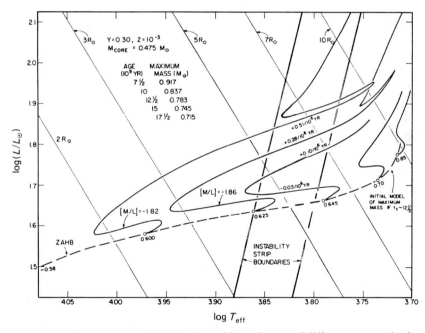

Fig. 14.10. Evolutionary tracks for horizontal branch stars of different masses in the luminosity, T_{eff} plane. When the horizontal branch stars develop a carbon core and the helium and hydrogen burning shell sources burn outwards, the stars increase slightly in luminosity and expand again, moving towards the red giant branch in the HR diagram. They populate the so-called asymptotic giant branch. Adapted from Iben (1971).

branch. We call such stars *asymptotic* giant branch stars. For population I stars this branch agrees with the first ascent red giant branch. The name originates from the metal poor globular clusters (see Figs. 1.7 and 1.8).

The population II horizontal branch stars also develop a growing carbon core, with a helium burning shell source around it. When the helium burning shell sources for these stars burn outwards the horizontal branch star envelopes expand again, becoming cooler but more luminous when the outer convection zone contributes to the energy transport outwards. Fig. 14.10 shows evolutionary tracks for horizontal branch stars of different masses. These stars evolve towards the first ascent giant branch but do not quite reach it. They remain slightly brighter than the first ascent giants, which is why this branch is called the asymptotic giant branch (see Figs. 1.7

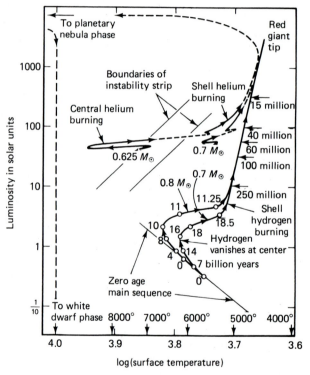

Fig. 14.11. Evolutionary tracks for population II stars with 0.7 and 0.8 $M_\odot$. A helium abundance of $Y = 0.30$ and a heavy element abundance of $Z = 10^{-3}$ was used. On the main sequence and subgiant branches evolution times since arrival on the zero age main sequence are given in billions of years. On the red giant branch evolution times from one arrow to the next are given in millions of years. Also indicated is the instability strip, the T_{eff}, L domain in which stars start to pulsate (see Chapter 17). From the tip of the asymptotic giant branch the stars probably evolve through the planetary nebula stage to become white dwarfs as indicated by the long dashed lines. From Iben (1971).

and 1.8, where the asymptotic giant branch can clearly be recognized). During their ascent on the asymptotic giant branch these stars have two shell sources.

Fig. 14.11 shows the whole evolutionary tracks for a 0.8 and 0.7 $M_\odot$ star as derived by Iben (1971). In this diagram evolutionary times for the different phases are also given.

The bluest horizontal branch stars with *very small envelope masses* do not evolve towards the asymptotic giant branch; instead, they evolve to the blue side of the HR diagram, towards the white dwarf region (see Chapter 16).

As seen in Fig. 14.12, which shows a composite color magnitude diagram for several globular clusters, the asymptotic and the red giant branches have tips. The stars do not evolve up the red giant branches any further. For metal rich stars (as in the globular cluster M67 or NGC 188) this tip occurs at lower luminosities than for metal poor stars like those in the globular cluster M92. Something must be happening at the tips of the asymptotic giant branches.

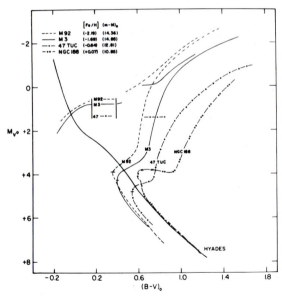

Fig. 14.12. A composite color magnitude diagram is shown for several star clusters with different heavy element abundances. [Fe/H] stands for log (Fe/H) − log (Fe/H)$_\odot$. The distance moduli for the different clusters were determined in such a way as to fit the main sequences with the luminosities indicated by the nearby metal poor stars of similar abundances (see Fig. 10.9). For the more metal rich stars the red giant branches terminate at lower luminosities than for the metal poor clusters. From Sandage (1986).

14.7   Planetary nebulae

We think we can see what is happening at the tip of the red giant branch by observing planetary nebulae (see Figs. 14.13 and 14.14; see also Volume 1). These planetary nebulae, which are bright mainly in the spectral emission lines, show a splitting of the line profiles (see Fig. 14.15). When the spectrograph slit extends over the diameter of the nebula, close to the edges of the nebula the line is seen as one, while in the center of the nebula there are two components, one shifted to the red and one to the blue, indicating that the nebula is expanding (see Fig. 14.16). Since the nebula is optically thin we see both the back side and the front side material. The front side material is moving towards us, the back side away from us, giving rise to the double line in the center. At the 'top' and the 'bottom' of the nebula the motion is perpendicular to the line of sight and no Doppler effect is seen. The expansion velocities are about 30 to 60 km s^{-1}. These nebulae expand from a central star which is observable but usually quite faint because most planetary nebulae are at vast distances. The nice thing about planetary nebulae is the fact that they have a 'clock' attached to them. Since they expand from a central star they increase their size in time. The larger the radius, the longer the time since

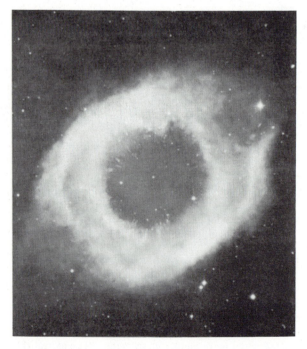

Fig. 14.13. The planetary nebula NGC 7293. (Photograph: Hale Observatories.)

the expansion or explosion started. We can thus study the time evolution of the nebula and with it the evolution of the central star. As discovered first by O'Dell (1968), the central stars are hotter for planetary nebulae with larger radii while the luminosities of the central stars remain nearly unchanged until they reach temperatures around 60 000 K. Once the central stars have reached such high temperatures (and small radii) their luminosities decrease. By this time they appear way below the hydrogen star main sequence, at a location where only helium stars or stars with still higher μ can be found. For these stars the nebulae are so large and have become so faint that they are very hard to detect and observations therefore stop.

Even in the earliest stages of evolution when the planetary nebula is still small the central stars of planetary nebulae are all hotter than 10 000 K, i.e. hotter than A0 stars. For these stars the planetary nebulae have already reached about half the maximum radius possible to be observed before they become too faint. We do not see the origin of the expansion, because the hydrogen gas of the planetary nebula needs a high temperature central light source to ionize it. Only after ionization does it shine in the emission

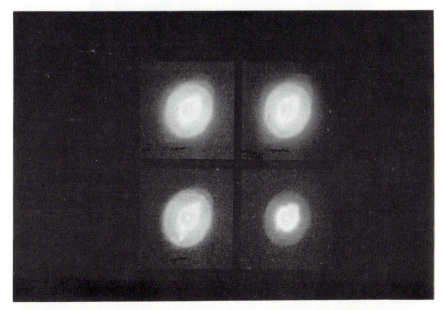

Fig. 14.14. Four photographs of the planetary nebula NGC 3242 taken with filters which are transparent only to light near the hydrogen Hα line at $\lambda = 6562.8$ Å (top left), near the forbidden O III line at $\lambda = 5006.8$ Å (top right), and near the forbidden N II line at $\lambda = 6583$ Å (bottom left). The He II line is emitted only in the hot inner regions of the nebula. For this line the nebula therefore appears smaller (bottom right). (Photograph: courtesy of B. Balick.)

lines we observe. As long as the central star is cooler than about 10 000 K the hydrogen gas in the planetary nebula is not ionized and the planetary nebula is dark.

We can now, as O'Dell first did, plot the position of the central stars of different planetary nebulae with different ages in the HR diagram (see Fig. 14.17). They form a sequence across it which we can interpret as a time or evolutionary sequence because we know that the coolest central stars are

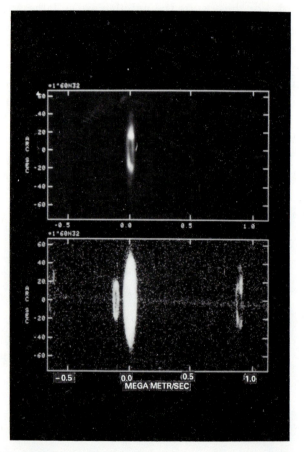

Fig. 14.15. High resolution spectra of the planetary nebula NGC 3242. In the top spectrum the brightness is shown as a function of velocity according to the relation $v = c\Delta\lambda/\lambda$. The velocity is given in units of 1000 km s^{-1}. The spectrograph slit was positioned along the 'long axis' of the nebula. In the bottom spectrum, the same spectrum is seen but the brightness shown is proportional to the logarithm of the flux received. This brings out the fainter lines. The lines seen are from left to right: He II at 6560.1 Å, Hα at 6562.8 Å, forbidden N II at 6583 Å. In the center the continuous spectrum of the faint central star is seen. All lines show splitting in central parts due to the expansion of the nebula as explained in Fig. 14.16. (Photograph: courtesy B. Balick.)

associated with the smallest and therefore youngest nebulae and the hottest ones with the largest old nebulae. The evolutionary arrow must go from right to left. We can also determine the time t for this evolution by measuring the radius increase of the nebula $\Delta r = v \times t$ and dividing it by the expansion velocity v. This tells us that the whole evolution of the observed planetary nebulae takes only about 20 000 years!

We can extrapolate this evolutionary sequence back in order to bridge the time when the nebula is dark and invisible. It seems reasonable to assume that the central star was originally at the tip of the red giant branch when it became unstable to a large mass loss and expelled about 10 to 50 per cent of its mass. Thus the whole hydrogen envelope was lost and expanded to become the planetary nebula while the central star must be mainly the carbon core of a red giant with a helium envelope and originally a helium burning shell source. If any hydrogen was left the hydrogen shell source probably extinguished because it got so close to the surface that it became too cool. We do not yet understand the reason for the expulsion of the envelope (radiative acceleration may be one cause), nor have we observed the process of formation of such a planetary nebula, simply because the nebula is cold and therefore invisible. Infrared observations may be able to detect planetary nebulae at a very early stage and several

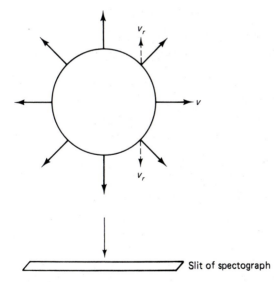

Fig. 14.16. Explains the splitting of the spectral emission lines of planetary nebulae. We see two parts of the expanding nebula one moving away from us (supposedly on the back side of the nebula) and one moving towards us (supposedly on the front side of the nebula). On the edges of the nebula the material is moving perpendicular to the line of sight and the line appears as one line, with a shift corresponding to the system velocity.

candidates have been identified. For some luminous cool supergiants in a binary system like α Her we see indications in the Ca K line profiles of some mass outflow. Whether this could be related to a later formation of a planetary nebula is not clear, nor do we understand the mechanism which leads to this outflow.

In Chapter 16 we will discuss what we think happens later to the central stars of planetary nebulae.

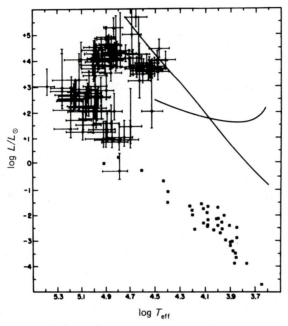

Fig. 14.17. The positions of central stars of planetary nebulae in the color magnitude diagram. The error bars are shown. While the nebula expands, the central star becomes smaller and hotter until it reaches the white dwarf region. The evolution goes from right to left. The main sequence and the horizontal branch positions are also shown by the solid lines. From O'Dell (1968).

15

Evolution of massive stars

15.1 Evolution along the giant branch

Just as for low mass stars, the evolution of high mass stars is caused by the change in chemical composition when hydrogen fuses to helium. These stars, however, have a convective core such that the newly formed helium is evenly mixed throughout the core. When hydrogen is consumed, the convective core contracts and also shrinks in mass (because the $\kappa + \sigma$ per gram decreases and therefore ∇_r decreases); the mixing then occurs over a smaller mass fraction, while some material, which was originally part of the convective region, is left in a stable region but with a slightly enriched helium abundance and also a slight increase in the N^{14}/C^{12} and C^{13}/C^{12} ratios. (See Figs. 13.2 and 13.4.) When the convective core mass reduces further, another region with still higher helium abundance and higher N^{14}/C^{12} and C^{13}/C^{12} is left outside the convection zone. The remaining convective core becomes hydrogen exhausted homogeneously while it contracts to a smaller volume and becomes hotter. The stars also develop hydrogen burning shell sources around the helium core. Again the core acts like a helium star with a very high temperature; the temperature at the bottom of the hydrogen envelope becomes too high to sustain hydrostatic equilibrium in the hydrogen envelope. The envelope expands and the stellar surface becomes cooler, moving the star in the HR diagram towards the red giant region. Again an outer hydrogen convection zone develops and reaches into deeper and deeper layers. Finally it dredges up some of the material which was originally in the convective core when it included a rather large mass-fraction of the star.

As we saw earlier, in massive stars the central gas pressures and densities are lower than in low mass stars; even during core contraction the densities in evolved high mass stars do not become high enough for electrons to degenerate. Therefore when the central temperature reaches 10^8 K helium fusion to C^{12} (i.e. the triple-alpha reaction) can start smoothly. The temperature rises, the pressure increases. The star expands somewhat and reaches a new equilibrium state.

For very massive stars ($M > 15\,M_\odot$) the interior temperatures become so high that helium burning starts while the stars are still close to the main sequence.

15.2   Blue loop excursions

Unlike the evolution of low mass stars, more massive stars do not increase their luminosities by large factors (see Fig. 15.1). Helium burning starts in the center when the helium core contains more than about $0.47\,M_\odot$. This happens when the stars reach the red giant region. The core then expands, as does the hydrogen burning shell source, thereby contributing less than before to the total energy generation. The temperature at the base of the hydrogen envelope decreases, the hydrogen envelope

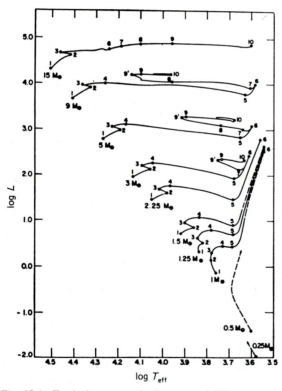

Fig. 15.1. Evolutionary tracks for stars of different masses. For more massive stars the luminosities of the red giants do not increase as much as those for lower mass stars. For the more massive stars the triple-alpha reaction starts soon after they reach the red giant region. The points with the numbers indicate the position for the onset of helium burning. The other numbers indicate other stages of evolution. From Iben (1967).

contracts and T_{eff} increases. In the HR diagram the star makes a 'blue loop excursion'. When more helium in the center is consumed, fusing to C^{12}, the number of particles decreases and the core contracts while heating up. The hydrogen envelope expands again. In the color magnitude diagram the star moves to the red again. This process finally creates a carbon star in the center of the helium star. The carbon star becomes more massive when the helium burning shell source burns its way outwards. The envelope continues to expand and cool (see Figs. 15.1 and 15.2). For the massive stars, several blue loop excursions may happen when new nuclear reactions start (see Fig. 15.2). How far the star moves to the blue depends sensitively on the chemical abundances and their stratification throughout the star. The blue loops extend further to the blue for low abundances of heavy elements than for higher metal abundances. Similarly, they extend further to the blue for higher masses than for lower masses. For masses around 2 or 3 solar masses (depending on chemical abundances) they seem to disappear. The stars remain close to the red giant branch. Fig. 15.2

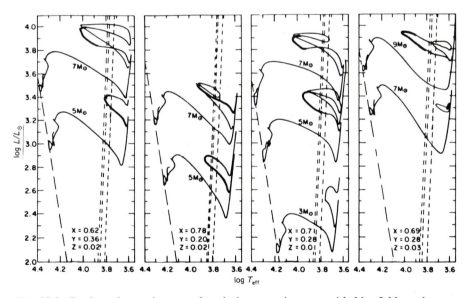

Fig. 15.2. During advanced stages of evolution, massive stars with $M > 3\,M_\odot$ make several blue loop excursions when new nuclear reactions become possible in their cores. The luminosities and the extent of the blue loops depend on the chemical composition, as may be seen from comparing the different panels. The abundances used are given in the panels. An increased helium abundance increases the luminosities. An increased abundance of heavy element decreases the luminosities and shortens the loops. The nearly vertical dashed lines show the Cepheid instability strip (see Chapter 17). Adapted from Becker, Iben and Tuggle (1977).

shows evolutionary tracks for stars of different masses and chemical composition as calculated by Becker, Iben and Tuggle (1977).

In Chapter 8 we saw that for the triple-alpha process the amount of energy liberated per gram is about 10 per cent of what is liberated by hydrogen burning (see Table 8.1). Less helium is burned during this stage than the amount of hydrogen burned on the main sequence. Therefore the lifetime of the star on the blue loop is only a few per cent of the main sequence lifetime. The stars spend most of this time near the tip of the blue loops. Since the conversion from helium to carbon in the core is a relatively slow process, the evolution from the red to the blue is much slower than the crossing from the blue to the red when the shell source burns its way out. In Fig. 15.2 stages of slow evolution are indicated by thicker lines.

15.3   Dependence of evolution on interior mixing

We saw earlier that the degree of mixing of helium enriched material to the outer layers is rather uncertain because of semi-convection as well as the unknown amount of convective overshoot and because of other possible mixing mechanisms like mixing due to strong differential rotation (called Schubert–Goldreich instability). It seems therefore important to study which changes in stellar evolution would be expected if the degree of mixing is larger than assumed in the calculations discussed so far.

In Fig. 15.3 we compare 'standard' evolutionary tracks ($\lambda = 0$) calculated in the standard way (which means assuming no overshoot mixing outside the convective core and assuming abundance stratification in the semi-convection zone which will just stabilize it) with evolutionary tracks calculated by Bertelli, Bressan and Chiosi (1985), in which they assume that due to convective overshoot the mixing extends to $\frac{1}{2}$ ($\lambda = 0.5$) or 1 ($\lambda = 1$) pressure scale height above the convectively unstable region. There are several striking differences between these tracks. For increased mixing we find the following:

(a) The main sequence lifetime of a star with a given mass becomes longer because a larger reservoir of hydrogen can be tapped before hydrogen is exhausted and the shell source develops.

(b) The luminosity L of the giant branch is increased relative to the main sequence stars of a given mass.

(c) The luminosity of the blue loops is increased even more than the luminosity of the giant branch such that there is now a larger difference in L between the blue loop star of a given mass and a giant of the same mass and also between a blue loop star and a main sequence star of equal masses.

For stars with increased mixing, the mass of an evolved star with given luminosity is therefore smaller than for a star which follows standard evolution theory. For overshoot mixing to about 1 scale height above the convective core the expected mass would be lower by about 20 per cent. The reduction in mass increases with the extent of overshoot.

If we determine the mass of blue loop stars from binaries we can determine the degree of additional mixing in stars. Present studies yield masses of supergiants smaller than expected from standard evolution theory, perhaps indicating additional mixing in the interiors of massive main sequence stars, but perhaps other corrections of our theory are also needed, such as changes in the interior absorption coefficients, for instance.

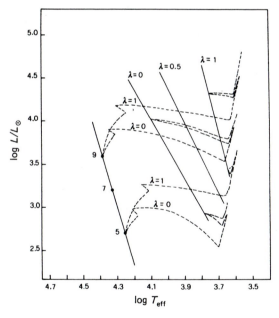

Fig. 15.3. Compares evolutionary tracks of stars with $M = 5\,M_\odot$ and $M = 9\,M_\odot$ for the same chemical composition but with different degrees of mixing during the main sequence phase. With more mixing ($\lambda = 1$ means overshoot by 1 pressure scale height) in the core of main sequence stars, the luminosity of the giant phase is increased as compared to no overshoot mixing ($\lambda = 0$). The luminosity for the blue loop is further increased. The solid lines on the right connect the tips of the blue loops for the different mixing parameters λ. The solid line on the left shows the main sequence. From Bertelli, Bressan and Chiosi (1984).

15.4   Evolution after helium core burning

In Fig. 15.4 we show the structure of a star after a large fraction of the helium star has been converted to carbon or perhaps oxygen depending on the central temperature which means, depending on the mass of the star. For higher temperatures C^{12} can combine with a He^4 nucleus to form O^{16}. In the core we therefore have a mixture of carbon and oxygen, surrounded by the remainder of the helium star, enclosed by the hydrogen envelope. At each boundary there is a shell source. Calculations show that at this stage of evolution the hydrogen shell source again becomes more important and the hydrogen envelope expands, the star evolves again to the red while the carbon oxygen core contracts and heats. Finally the central temperature increases enough to permit further nuclear reactions. Two C^{12} may combine to Mg^{24}. The star may experience a new blue loop, but the lifetime on this blue loop is still shorter because less energy is gained in this burning process. We are less likely to see stars on the second blue loop. For any supergiant, we are most likely to see it on the first blue loop, probably crossing from the red to the blue. In a few cases it may cross in the other direction. Most of the cool supergiants must have been red giants at least once before.

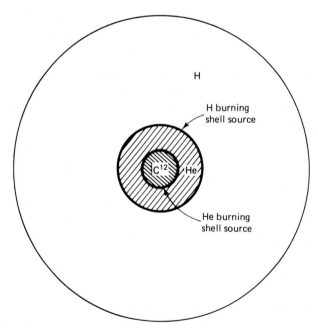

Fig. 15.4. Shows schematically the structure of a star after a large fraction of the core helium star has been converted to C^{12}. For higher temperatures some of the C^{12} may be converted to O^{16}.

15.5   The carbon flash

For massive stars we do not expect a helium flash because the densities in the helium core are not high enough for electron degeneracy. When, however, the star converts helium to carbon, the core contracts further. For stars with masses between 2.25 $M_\odot$ and about 5 $M_\odot$ the density may after helium burning become high enough for electron degeneracy. The onset of C^{12} burning then leads to a runaway energy generation until the temperature becomes high enough to remove the degeneracy and the core expands such that the nuclear reaction rate decreases and a new equilibrium can be established. It is not yet clear whether an explosive onset of carbon burning could perhaps lead to an explosion of the star. This depends on the number of neutrinos generated at high temperatures. Since they can freely escape they provide an efficient cooling mechanism which slows down the runaway heating. It would be interesting to find out whether the explosive onset of carbon burning could indeed lead to a supernova explosion.

15.6   Evolution of massive stars beyond the blue loops

It is not difficult to extrapolate further what will happen in the interior of massive stars after carbon is exhausted in the core and the core further contracts and heats. Nuclear reactions between particles with increasing Z can take place, building up heavier and heavier elements, until elements of the iron group like Ni, Fe, Co are formed. Up to these nuclei such fusion reactions still liberate energy because the mass per nucleus still decreases. A binding energy of Δmc^2 is liberated when particles combine (see Table 8.1). Nuclear build-up will, however, not proceed beyond the iron group nuclei because energy would be consumed, not liberated, in building up heavier elements. We end up with a star which qualitatively looks as seen in Fig. 15.5. The star has a nucleus with iron group elements surrounded with shells of lighter elements (like onion layers). The outermost shells are helium and hydrogen. Fig. 15.5 applies to stars with masses greater than around 12 $M_\odot$ to 15 $M_\odot$.

15.7   Type II supernovae

Type II supernovae probably do not occur in old stellar systems like elliptical galaxies. It is therefore believed that they are associated only with young stellar populations. They are often associated with large H II regions, that is with regions where hydrogen is ionized. This means they must be associated with hot stars which can provide enough energetic photons for the ionization. These observations suggest that they may be related to very advanced stages of evolution of rather massive stars. The large amount of energy involved in this explosion ($\sim 10^{52}$ erg) leaves only two possibilities for the energy supply: either nuclear reactions or the formation of a neutron star (see Section 8.2 and Volume 1). In the formation of a neutron star of 1 $M_\odot$ with a radius of $\sim 10^6$ cm (~ 10 km) the gravitational energy release E_g is

$$E_g = \frac{GM^2}{R} \sim \frac{6.6 \times 10^{-8} \times 4 \times 10^{66}}{10^6} \sim 30 \times 10^{52} \text{ erg}$$

plenty of energy for the supernova explosion. The total nuclear energy E_n available is about ΔMC^2, where ΔM is the mass fraction converted into energy, about 1 per cent of the stellar mass. For a 10 solar mass star we thus find

$$E_n \sim 10^{-2} \times Mc^2 = 10^{-2} \times 2 \times 10^{34} \times 9 \times 10^{20} \text{ erg} = 2 \times 10^{53} \text{ erg}$$

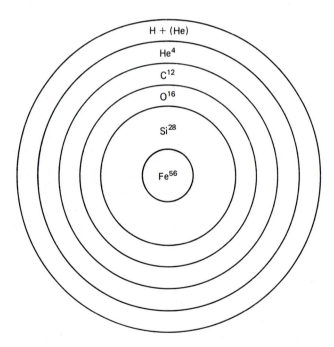

Fig. 15.5. Very massive stars ($M > 12\ M_\odot$) will manufacture heavier and heavier elements in their interior during advanced stages of stellar evolution. The heaviest nuclei are found in the innermost core, which is surrounded by shells with successively lighter elements. The star looks like an onion with different shells (not drawn to scale). According to Clayton (1968).

which would also be enough for the supernova explosion. On the other hand if the supernova is related to *late* stages of evolution for massive stars then essentially all of the nuclear *energy has been used already* to provide the luminosity of the star during its lifetime.

Once the star with 10 to 40 $M_\odot$ has reached the stage shown in Fig. 15.5 a further increase in central density and temperature due to the nuclear reactions outside the Fe^{56} core will not lead to more nuclear energy generation in the interior but instead a process analogous to 'ionization' occurs. The Fe^{56} breaks up into helium by the following photo disintegration process:

$$Fe^{56} \rightarrow 13He^4 + 4n - 124\,MeV$$

where n stands for neutron. This means a large energy drain occurs for the central region. (Since the pressure is provided by the free electrons this process does not increase the pressure because of the increasing number of heavy particles.)

At these temperatures and densities still another process takes place. The protons in the nuclei combine with electrons to form neutrons by the process

$$p + e^- \rightarrow n + \nu_e$$

Electrons and positrons may also annihilate to create neutrinos. These processes reduce the pressure because electrons are consumed and energy is lost. Because of the reduced electron pressure, which now cannot support the weight of the overlying material, the Fe core collapses almost freely until nuclear densities are reached. In the core essentially all electrons and protons combine to neutrons at such high densities. At that point resistance to further compression increases steeply and the collapse has to stop. The pressure change causes sound waves which are trapped and build up to form a shockwave which for stellar masses between about 10 and 15 $M_\odot$ can apparently lead to an explosion of the outer layers of the star. The theoretical result is the formation of a neutron star in the center with a mass which is nearly equal to the original iron core mass but which now consists of neutrons. The break-up of Fe consumes about two-thirds of the energy released in the collapse but just enough energy seems to be left for the explosion. For stars more massive than about 15 or 20 $M_\odot$ no explosion seems to result theoretically so far. The whole star must then collapse forever. Very massive stars may have no other choice than ultimately to become black holes. Of course, for us onlookers this takes an infinite amount of time.

16

Late stages of stellar evolution

16.1 Completely degenerate stars, white dwarfs

In Chapter 14 we saw that low mass stars apparently lose their hydrogen envelope when they reach the tip of the asymptotic giant branch. What is left is a degenerate carbon–oxygen core surrounded by a helium envelope. The mass of this remnant is approximately 0.5 to 0.7 solar masses depending perhaps slightly on the original mass and metal abundances. The density is so high that the electrons are partly or completely degenerate except in the outer envelope. We also saw that central stars of planetary nebulae seem to outline the evolutionary track of these remnants which decrease in radius, still losing mass and increasing their surface temperature. Their luminosities do not seem to change much until they reach the region below the main sequence (see Fig. 14.14). In the interiors these remnants are not hot enough to start any new nuclear reactions. When they started to lose their hydrogen envelope they still had a helium burning and a hydrogen burning shell source. When the hydrogen envelope is lost the hydrogen burning shell source comes so close to the surface that it soon becomes too cool and is extinguished. The helium burning shell source survives longer but finally is also extinguished, when the star gets close to the white dwarf region. The remnant ends up as a degenerate star with no nuclear energy source in its interior but which still has very high temperatures. This is the beginning of the evolution of a white dwarf. It loses energy at the surface, which is replenished by energy from the interior, i.e. by thermal energy from the heavy particles. The electrons are completely degenerate and cannot reduce their energy. Because of the energy loss the star must slowly cool down. Unlike the situation for non-degenerate stars, the pressure does not change when a degenerate star cools since the star is balanced by the degenerate electron pressure which does not change with decreasing temperature. Therefore the star does not contract; it just cools off, maintaining its size.

Where are these stars found in the HR diagram? For hydrostatic equilibrium we must again require

$$\frac{dP_g}{dr} = -\rho \frac{GM_r}{r^2} \quad \text{with} \quad \frac{dM_r}{dr} = 4\pi r^2 \rho \qquad (16.1)$$

The difference with respect to stars considered so far is the relation between P and ρ, which is independent of the temperature for completely degenerate stars. To integrate equations (16.1) we therefore do not need any equations describing the temperature stratification as we did for non-degenerate stars. As we saw in Section 14.3, we have $P_g = K_1(\rho/\mu_E)^{5/3}$ provided we are not dealing with relativistic degeneracy. Here μ_E is the mass (of heavy particles) per degenerate particle, in this case per electron. $K_1 = 9.9 \times 10^{12}$ cgs units. Equations (16.1) are then two differential equations for the two unknown functions $M_r(r)$ and $P_g(r)$ with $\rho = \rho(P_g)$ (or $P_g = P_g(\rho)$) and with two boundary conditions: one at the center, namely at $r = 0$, $M_r = 0$ and one at the surface, namely $M_r = M$ for $P_g = 0$. For given M we find *one* solution for $M_r(r)$ and $P_g(r)$ with one set of values for the central pressure P_c and for R. The solution does not depend on T_{eff} or L, which remain undetermined by the hydrostatic equation. A whole series of solutions with different T_{eff} and L is possible for a given M. **For degenerate stars the hydrostatic equilibrium alone determines only R or ρ, not the central temperature T_c.**

In order to see qualitatively how M and R are related we proceed in a similar way as we did in Chapter 3 when we derived a preliminary mass–luminosity relation for main sequence stars. Since the temperature and therefore the luminosity does not enter here we expect to find a relation between mass and radius only. We can derive a qualitative relation between M and R if we replace all the variables by average values. We can say

$$\overline{\frac{dP_g}{dr}} \approx -\frac{P_c}{R}, \quad \bar{\rho} = \frac{M}{\frac{4}{3}\pi R^3} \quad \text{and} \quad \overline{\frac{M_r}{r^2}} \sim \frac{M}{R^2} \qquad (16.2)$$

With this we find from equation (16.1) that

$$\frac{P_c}{R} \sim \frac{GM}{R^2} \frac{3}{4\pi} \frac{M}{R^3} = \frac{GM^2}{R^5} \frac{3}{4\pi} \qquad (16.3)$$

From the equation of state for a completely degenerate gas (equation 14.16) we have

$$P_g \sim K_1 \frac{\rho^{5/3}}{\mu_E^{5/3}} \sim \frac{M^{5/3}}{R^5} \frac{K_1}{(\frac{4}{3}\pi)^{5/3}} \frac{1}{\mu_E^{5/3}} \qquad (16.4)$$

and

$$\frac{P_c}{R} \sim \frac{M^{5/3}}{R^6} \frac{K_1}{(\frac{4}{3}\pi)^{5/3}} \frac{1}{\mu_E^{5/3}} \tag{16.5}$$

The hydrostatic equation (16.3) now reads

$$\frac{1}{(\frac{4}{3}\pi)^{5/3}} \frac{K_1}{\mu_E^{5/3}} \frac{M^{5/3}}{R^6} \sim \frac{GM^2}{R^5} \frac{1}{\frac{4}{3}\pi} \tag{16.6}$$

Solving for the radius we find

$$R \sim M^{-1/3} \frac{1}{\mu_E^{5/3}} \frac{K_1}{G} \frac{1}{(\frac{4}{3}\pi)^{2/3}} \tag{16.7}$$

For a given chemical composition the radius of the stellar remnant *decreases* with *increasing mass as* $M^{-1/3}$. For increasing masses, the gravitational forces increase; therefore the pressure forces must also increase. These can only increase if the density ρ increases, which means the star must become smaller. The situation is very different from the one for main sequence stars where larger mass stars have larger radii due to the higher temperature in the massive stars.

How large are the radii estimated from the rough approximation (equation (16.7))?

For a fully ionized carbon star we have 6 electrons for 12 nuclei. The mass μ_E per electron is $\frac{12}{6}m_H = 2m_H = 3.32 \times 10^{-24}$ g. We found $K_1 \sim 10^{13}$ (equation (14.17)). We then obtain

$$R \sim M^{-1/3} \frac{1}{\mu_E} \frac{10^{13}}{6.7 \times 10^{-8}} \left(\frac{3}{4\pi}\right)^{2/3} \sim M^{1/3} \times 1.82 \times 10^{19} = 1.44 \times 10^8 \text{ cm}$$

for $M = M_\odot = 2 \times 10^{33}$ g. For a one solar mass star we thus estimate $R \sim 1400$ km. This is the correct order of magnitude for a white dwarf's radius, in spite of our very rough approximations.

In Volume 1 we estimated from the luminosity of a white dwarf at solar temperature that its radius is about 6000 km. With our crude estimate we are wrong by a factor of 4, which is quite good, considering that the radius of a white dwarf is about 100 times smaller than that of the sun. White dwarfs must then be degenerate stars. **For white dwarfs we must therefore have $R \propto M^{-1/3}$ for non-relativistic degeneracy.** (For $M > 0.2\, M_\odot$ corrections are already needed.)

As we emphasized earlier, **the effective temperatures and luminosities are not determined by the hydrostatic equation.** The T_{eff} are independent of this condition. As for main sequence stars the surface temperature or effective temperature is determined by heat transport from the inside out. The difference is that the central temperature of a white dwarf is not determined by hydrostatic equilibrium but rather by its history, by whatever the central temperature was when the star arrived in the white dwarf region of the HR diagram, and by its age as a white dwarf. Since it started out as the very high temperature nucleus of a red giant we expect it to arrive as a hot star with $T_{\text{eff}} \sim 10^5$ K. Even with a radius of only 6000 km it will still be a fairly bright object intrinsically and loses energy rather rapidly. This reduces its temperature. It has no nuclear energy source, it does not contract because the pressure is independent of T, so it just cools down, becoming fainter. Its luminosity is $L = 4\pi R^2 \sigma T_{\text{eff}}^4$. For a given mass the radius R is fixed, while T_{eff} and L slowly decrease. In the $\log L$, $\log T_{\text{eff}}$ diagram it follows a line

$$\log L = 4 \log T_{\text{eff}} + 2 \log R(M)$$

R is a constant for a given mass but decreases with increasing mass. In the $\log L$, $\log T_{\text{eff}}$ diagram the cooling sequence of a white dwarf is a straight line, the larger mass white dwarfs having the smaller L for a given T_{eff}, as shown in Fig. 16.1.

For more massive white dwarfs the radius becomes smaller and smaller. How small can white dwarfs get? For larger masses and smaller radii the density obviously increases, and with increasing density the degeneracy increases. For large enough masses relativistic degeneracy is approached. The electrons must have velocities close to the velocity of light. In Chapter 14 we derived that for relativistic degeneracy

$$P_g = K_2 \left(\frac{\rho}{\mu_{\text{E}}}\right)^{4/3} \tag{16.8}$$

If we want to find out whether there is a limiting radius of white dwarfs for increasing masses we have to use this relation which holds for extremely large densities.

Using the same approach as above, we still find equation (16.3) but now equation (16.8) has to be used to replace P_g. We find

$$\frac{P_g}{R} \sim K_2 \frac{M^{4/3}}{R^5} \frac{1}{(\frac{4}{3}\pi)^{4/3}} \approx \frac{GM^2}{R^5} \frac{3}{4\pi} \tag{16.9}$$

We cannot solve for the radius, because R cancels out of this equation. We are left with an equation for the mass alone, which becomes

$$M^{2/3} \sim \frac{K_2}{G}\left(\frac{3}{4\pi}\right)^{1/3}\frac{1}{\mu_E^{4/3}}$$

and

$$M \sim \left(\frac{K_2}{G}\right)^{3/2}\left(\frac{3}{4\pi}\right)^{1/2}\frac{1}{\mu_E^2} = M_e \qquad (16.10)$$

With $\mu_E = 2m_H$ as for a helium, carbon or oxygen white dwarf we estimate $M_e \sim 1.4\,M_\odot$.

Equation (16.10) tells us that **there can be only one mass for a relativistic degenerate star in hydrostatic equilibrium and that mass is $M_e \approx 1.4\,M_\odot$**, as was first calculated by Chandrasehkar. We do not obtain any information about the radius. The star could have any radius, provided it is still a star whose pressure is determined by relativistic degenerate electrons, which requires very small radii. In the hydrostatic equation the pressure cancels out.

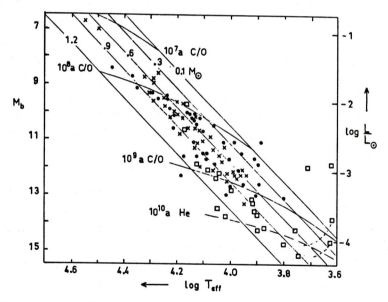

Fig. 16.1. In the M_{bol}, T_{eff} diagram cooling sequences for white dwarfs with $1.2 \geqslant M/M_\odot \geqslant 0.1$ are shown (solid lines). They are straight lines going almost diagonally through the diagram. The mass numbers are given at the top. The curved solid lines show the positions of white dwarfs of different ages. A carbon–oxygen core with a thin hydrogen or helium shell was assumed. The positions of helium white dwarfs with an age of 10^{10} years are also shown by the dashed line. In addition the positions determined for existing white dwarfs have been plotted. Different symbols refer to different spectral types of white dwarfs. From Weidemann (1975).

Complete relativistic degeneracy is of course only a limiting case which is never quite reached. The derivation shows, however, that there cannot be a white dwarf with $M > M_e$.

If μ_E changes, the mass limit changes and a relativistically degenerate star with $M = 1.4\,M_\odot$ is not in hydrostatic equilibrium if $\mu_E \neq 2$.

16.2 Neutron stars

We saw that for very large electron densities leading to relativistic degeneracy we reach a mass limit for white dwarfs. For larger masses gravitational forces are larger than the pressure forces. The star shrinks.

We know, however, from our discussion of supernovae that for very high densities the electrons and protons are squeezed so close together that they form neutrons. We then no longer have degenerate electrons, but degenerate neutrons. Because of the larger mass of neutrons as compared to electrons the degeneracy starts only at much higher densities as we discussed in Chapter 14, but for sufficiently high densities we also find neutron degeneracy. The constant K_1 in relation (16.4) is smaller for a neutron star because m_e for the electrons has to be replaced by m_n, the larger mass of the neutrons. Also instead of μ_E, the mass per degenerate electron (which is $\mu_E = 2m_H$ for helium and carbon white dwarfs), we now have to insert μ_N, the mass per degenerate neutron, which gives $\mu_N = m_H$ if we have only neutrons. There are also some heavier particles in such high density matter, which increases μ_N but by how much we do not know exactly. In any case $\mu_N = m_H$ is a reasonable estimate. This means we have less mass per degenerate particle.

We again find that the radii of neutron stars decrease as $M^{-1/3}$ for non-relativistic degeneracy and again we can estimate a radius for a $1\,M_\odot$ neutron star, which is decreased by a factor 1835 because of the larger neutron mass, but increased by the factor $2^{5/3}$ because of the smaller value of μ_N. Altogether we find a decrease in radius by roughly a factor of 600 as compared to the white dwarf, that is a $1\,M_\odot$ neutron star has a radius of about 10 to 15 km. The density in such a star comes out to be

$$\bar\rho \sim \frac{2 \times 10^{33}}{(\tfrac{4}{3}\pi) \times 10^{18}}\ \text{g cm}^{-3} \sim 5 \times 10^{14}\ \text{g cm}^{-3}$$

One cubic centimetre of this material has a mass of several 100 million tonnes. In other words, such a star has nuclear densities.

Is there a limiting mass for neutron stars? The constant K_2 does not depend on the particle mass. It is therefore the same for a relativistically degenerate neutron star as for a star with relativistically degenerate electrons. The limiting mass M for a neutron star, which we call M_N, is proportional to μ_N^{-2} as seen from equation (16.10). If $\mu_N = m_H$ the limiting value $M_N \sim 4M_e \sim 5.6\ M_\odot$. In reality we have to consider the presence of heavier nuclei, which increases μ_N. The upper limit for the mass of a neutron star appears to be between 3 and 5 solar masses.

The actual mass of a neutron star depends, of course, on the history of its formation. If it originates from the collapse of the Fe core of 1.35 solar masses, while the remainder of the stellar mass is expelled, we will find a supernova remnant which is a neutron star with $M \approx 1.35\ M_\odot$.

If a He or C, O white dwarf with a mass close to $1.4\ M_\odot$ accretes mass from a companion star it is squeezed together and heats up. He or C^{12} may then start nuclear reactions before the star is compressed into a neutron star. In the highly degenerate interior such nuclear reactions may lead to a detonation or deflagration which can be the origin for a supernova I explosion. These also occur in elliptical galaxies and are therefore believed to have old, low mass stars as progenitors.

17

Observational tests of stellar evolution theory

17.1 Color magnitude diagrams for globular clusters

The best way to check stellar evolution calculations is, of course, to compare calculated and observed evolutionary tracks. Unfortunately we cannot follow the evolution of one star through its lifetime, because our lifetime is too short – not even the lifetime of scientifically interested humanity is long enough. Only in rare cases may we observe changes in the appearance of one star, for instance when it becomes a supernova. Another example occurred some decades ago when FG Sagittae suddenly became far bluer, a rare example of stellar changes which are too fast to fit into our present understanding of stellar evolution.

Generally evolutionary changes of stars are expected to take place over times of at least 10^4 years (except perhaps for stars on the Hayashi track, where massive stars may evolve somewhat faster). How then can we compare evolutionary tracks? Fortunately there are star clusters which contain up to 10^5 stars all of which are nearly the same age but of different masses. In such very populous clusters there are a large number of stars which have nearly the same masses.

In Fig. 17.1 we show schematically evolutionary tracks of stars with about one solar mass. They all originate near spectral types G0 or G2 on the main sequence. Their lifetime, t, on the main sequence is about 10^{10} years. The evolution to the red giant branch takes about 10^7 years. So after 10^{10} years stars with a mass of $1\,M_\odot$ are just leaving the main sequence. Stars with a mass of $1.004\,M_\odot$ have a lifetime of $10^{10} - 10^7$ years (if $t \propto M^{-2.5}$). Stars with a mass of $1.004\,M_\odot$ therefore are further along in their evolution and have just reached the red giant branch. Stars with $1.008\,M_\odot$ are still further evolved and have already reached the top of the red giant branch. Since stars of $1.000\,M_\odot$, $1.004\,M_\odot$ and $1.008\,M_\odot$ all have indistinguishable evolutionary tracks we will find that they outline the evolutionary track for stars with $1.004 \pm 0.004\,M_\odot$.

This means, of course, that we must have a large number of stars in a given cluster to see enough stars along the evolutionary track. More accurately, what we see in an HR diagram is an isochrone showing stars at a given time, i.e. at a given age, if they were all 'born' at the same time. These isochrones are almost identical with an evolutionary track for a given stellar mass.

How well do these isochrones agree with stellar evolution theory? In Fig. 17.2 we compare isochrones calculated by VandenBerg and Bell (1985) with the observed color magnitude diagrams (isochrones) for 47 Tuc, M5 and the cluster Pal 12 (Pal stands for Palomar) as given by Stetson *et al.* (1989). For 47 Tuc and M5 the isochrones for an age of 16×10^9 years agree rather well with the observed positions of the stars. For the cluster Pal 12 the agreement for this age is rather poor; however, in Fig. 17.3 we see that for this cluster the isochrone for an age of 12×10^9 years give very good agreement. We can conclude that present theoretical model calculations can represent the observations quite well. Of course, we have some parameters to adjust, like age and the chemical abundances of heavy

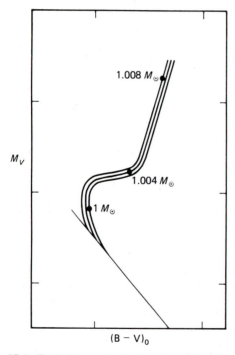

Fig. 17.1. Evolutionary tracks for stars with masses of 1 $M_\odot$, 1.004 $M_\odot$ and 1.008 $M_\odot$ are shown schematically. After 10^{10} years the star with a mass of 1.008 $M_\odot$ may have just reached the tip of the red giant branch while the star with 1.004 $M_\odot$ just arrives at the bottom of the red giant branch. The star with 1 $M_\odot$ is just leaving the main sequence.

elements which we only know from colors or spectral analysis to within a factor of 2 or 3. The helium abundance is also uncertain.

Theoretical model calculations yield L and T_{eff}. The observations give m_V and B − V. We therefore have to establish the relation between B − V, T_{eff} and the bolometric corrections which all depend on metal abundances on interstellar reddening, on distances and on the theory of stellar atmospheres.

Considering all the steps necessary before a comparison can be made we can be quite satisfied with the agreement between observed and calculated isochrones. The main sequence and giant branches can now be well represented. We therefore feel confident that basically our understanding of stellar evolution is correct.

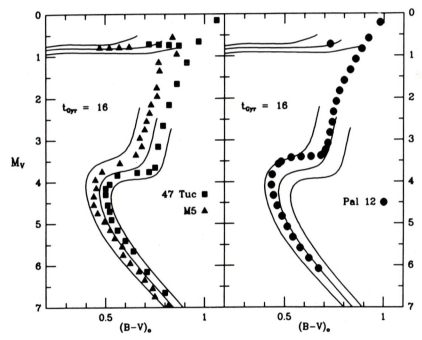

Fig. 17.2. Color magnitude diagrams of the globular clusters 47 Tuc and M5 are shown as measured by Hesser *et al.* (1987). Superimposed are theoretical isochrones for stars with ages of 16×10^9 years as calculated by VandenBerg and Bell (1985), assuming a helium abundance by mass $Y = 0.20$ and heavy element abundances $\log Z/Z_\odot = -0.49$, -0.79 and -1.27. For the convection a characteristic length $l = 1.5H$ was assumed. A pseudo-distance modulus $m_V - M_V = 13.15$ was adopted for 47 Tuc and $E(B - V) = 0.04$. For M5 a pseudo-distance modulus of $m_V - M_V = 14.15$ and $E(B - V) = 0.03$ was used. For M5 and 47 Tuc a good fit is found for an age of 16 billion years with $\log Z/Z_\odot = -1$ for M5 and $\log Z/Z_\odot = -0.65$ for 47 Tuc. For the cluster Pal 12, for which $m_V - M_V = 16.3$ and $E(B - V) = 0.02$ was adopted, no good fit is found for this age. From Stetson *et al.* (1989).

Generally ages between 12 and 17 billion years are obtained in this way for globular cluster stars. These are larger than the age of the universe as presently derived from the Hubble expansion. We do not yet understand the origin of this discrepancy.

17.2   Color magnitude diagrams of young clusters

As pointed out above we need clusters with a large number of stars in order to find enough stars in a very small mass range to outline the evolutionary track for a given stellar mass. The young galactic clusters do not have enough stars. In most young galactic clusters we find the main sequence and perhaps one or at most a handful of supergiants or giants. Galactic clusters are therefore not well suited for the study of evolutionary tracks.

Our best chances of studying evolutionary tracks of massive stars are provided by the populous young clusters in the Magellanic Clouds, though in these dense distant clusters the contamination by background stars causes a large scatter in the photometric data. In Fig. 17.4 we show a color

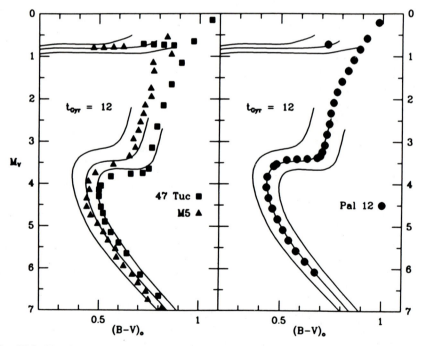

Fig. 17.3. For the cluster Pal 12 an age of 12 billion years and a heavy element abundance of $\log Z/Z_\odot = -0.8$ gives a good match for observed and theoretical isochrones in the color magnitude diagram. The metal abundances for the isochrones plotted are the same as those in Fig. 17.2. From Stetson *et al.* (1989).

magnitude diagram for the LMC cluster NGC 2010 as measured by Mateo (1987). Evolutionary tracks are also shown. The scatter of the data is large, but there is a fair overall agreement between theoretical and observed isochrones. In Fig. 17.5 we show a comparison of theoretical tracks with the color magnitude diagram of NGC 330 in the Small Magellanic Cloud. Fig. 17.6 shows a photograph of this young populous cluster. Generally there is reasonable agreement between the observed color magnitude diagram and the theoretical isochrones except for a shift in B − V. The color excess for this cluster may be larger than the assumed value of $E(B − V) = 0.06$, generally adopted for the Small Magellanic Cloud galaxy.

17.3   Observed masses of white dwarfs

Our theoretical discussions have shown that white dwarfs must have masses $M < 1.4\, M_\odot$. The mass of a white dwarf must be determined

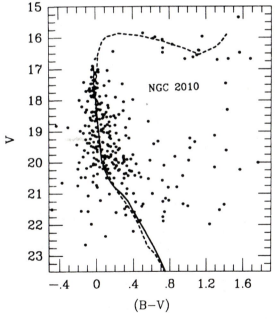

Fig. 17.4. The color magnitude diagram for the cluster NGC 2010 in the Large Magellanic Cloud is shown as measured by Mateo (1987). Superimposed are isochrones as calculated by Brunish and Truran (1982) for an age of 63 million years (solid line) and by Bertelli, Bressan and Chiosi (1984) for 172 million years with convective overshoot (dashed line). A helium abundance Y = 0.28 and solar metal abundances were assumed. A color excess correction of $E(B − V) = 0.09$ was applied. A distance modulus of $m_{V0} − M_V = 18.20$ was adopted. From Mateo (1987).

by the history of its formation. It depends on the mass of the carbon–oxygen core in its progenitor red giant and on the mass of the helium envelope still remaining after the mass-losing process.

For some white dwarfs in binaries masses can be determined as, for instance, for Sirius B. Masses $\leqslant 1\ M_\odot$ are generally found (see Fig. 16.1), though Sirius B has a mass of $1.05\ M_\odot$.

Some of the white dwarfs are close enough for us to measure parallaxes, such that their luminosities can be determined. If a white dwarf has a main sequence companion the photometric parallax of the companion can be derived. The effective temperatures of the white dwarfs are obtained from the energy distributions. The stars can thus be placed in the $\log L$, $\log T_{\mathrm{eff}}$ diagram, and their positions can be compared with theoretical cooling tracks for different masses.

In Fig. 16.1 we compare the positions of the white dwarfs in the $\log L$, $\log T_{\mathrm{eff}}$ diagram with theoretical cooling sequences for different masses. The white dwarfs cluster along a sequence with $M = 0.6 \pm 0.2\ M_\odot$ with a few exceptions, like Sirius B. In dwarf novae (which are binaries believed

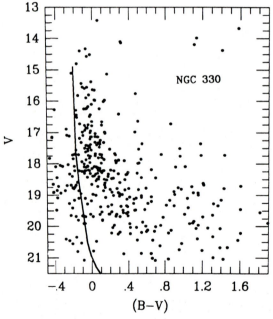

Fig. 17.5. The color magnitude diagram of the populous young cluster NGC 330 in the Small Magellanic Cloud (SMC) is shown. Superimposed is an isochrone for the age of 1 million years as calculated by Brunish and Truran (1982). $Y = 0.28 \log Z/Z_\odot = -1.3$ were assumed. A distance modulus of $m_V - M_V = 18.8$ was adopted for the SMC, and $E(B - V) = 0.06$. All stars appear to be too red. We suspect that the $E(B - V)$ for this cluster may be larger by 0.12 than assumed. From Mateo (1987).

to have a white dwarf that accretes mass from its companion) the masses of the white dwarfs appear to be frequently larger than $0.6\,M_\odot$ probably due to the mass accretion. In the past perhaps Sirius B also accreted mass from Sirius A?

The masses around $0.6\,M_\odot$ are consistent with the picture that white dwarfs are the burnt-out cores of asymptotic branch giants. By the time the star reaches the termination point of the asymptotic giant branch the core reaches a mass around $0.6\,M_\odot$. The masses of the central stars of planetary nebulae also appear to cluster around $M = 0.6\,M_\odot$. as expected for remnant cores of low mass stars.

While generally the white dwarf observations agree with our theoretical expectations there are still some problems. In a young galactic cluster, the Pleiades, we see stars with masses up to $6\,M_\odot$ still on the main sequence, yet there is at least one white dwarf in that cluster. Since only stars with masses larger than $6\,M_\odot$ have evolved off the main sequence the progenitor star for this white dwarf, which has certainly less than the limiting mass (namely 1.4 solar masses), must have expelled about $5\,M_\odot$ to become a

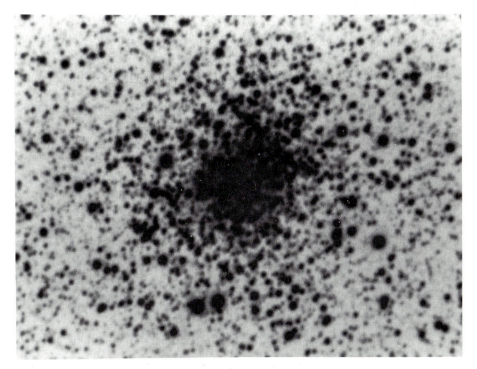

Fig. 17.6. A (negative) photograph of the young, populous cluster NGC 330 in the Small Magellanic Cloud galaxy. Courtesy: P. Hodge.

white dwarf. How did this happen? By the formation of a massive planetary nebula, perhaps?

17.4   Supernovae, neutron stars and black holes

17.4.1 Supernovae and neutron stars

Supernovae have always been of special interest to the astronomers because of the enormous amount of energy involved. The final decline of the light curves with a half-life of 56 days suggests that a large amount of Co^{56} is formed in the explosion, which decays radioactively with a half-life of 56 days and which probably supplies the energy during the slow decrease in light output. This in itself is not very strong evidence that we understand what is happening in such a gigantic explosion. Fortunately a supernova explosion took place in 1987 in the Large Magellanic Cloud, the nearest neighbor galaxy, and indeed a small number of high energy neutrinos were detected at several neutrino observing stations. These were detected a few hours before the light outburst, consistent with the picture that the core collapse produces many energetic neutrinos which escape with the velocity of light. The shock front leading to the explosion develops about 30 seconds later and travels outward with supersonic but much lower velocity than the speed of light. It reaches the surface a few hours later with high temperatures ($\sim 10^5$ K) when it causes the sudden brightening of the star. A very large neutrino flux must have occurred to lead to a detection of even a handful of neutrinos. This and the time delay between the neutrino burst and the optical brightening of the supernova give us confidence that our theoretical studies may be close to the truth.

Are neutron stars remnants of supernova explosions? Very rapidly rotating pulsars appear to be neutron stars for reasons discussed in Volume 1. Their pulse period is their rotation period, which makes a beam of light pass by us once or twice during one revolution.

A pulsar with a period of one-thirtieth of a second is seen in the Crab nebula, the remnant of a supernova recorded by Chinese astronomers in the year 1054. Another pulsar is seen in the Vela supernova remnant. There may be neutron stars in other supernova remnants which we cannot recognize as such because their light beams may be directed away from us. We do not know yet whether there are neutron stars in all remnants of type II supernovae.

The origin of type I supernovae is unknown at present, though it is speculated that mergers of white dwarfs or neutron stars are responsible.

17.4.2 Black holes

Are there black holes? The first question is: How could we recognize them? Large gravitational fields may be expected to draw nearby material into the black hole, especially in binaries. Such material would be accelerated to very high velocities. High velocity particles are expected to emit X-rays. Of course, all this has to happen outside the black hole, otherwise we could not see it. We are therefore looking for X-ray sources which are not neutron stars (X-ray generation can also be expected for mass accreting neutron stars). If the mass of the X-ray source is larger than about 5 $M_\odot$, the limiting mass for a neutron star, we can be sure it is not a neutron star. Mass determinations can only be done in binaries. For a black hole binary we can, of course, only see one star, the companion of the black hole. For one star we can only determine the $(M_1 + M_2) \sin^3 i$ (see Volume 1), where i is the (unknown) inclination between the line of sight and the normal to the orbital plane.

It is the uncertainty of the factor $\sin i$ which has prevented us from drawing any firm conclusion about the existence of a black hole in a binary system. There are, however, a few systems in our galaxy and in the Large Magellanic Cloud which are highly probable candidates. Massive black holes may be in the centers of quasars and other active nuclei of galaxies. While this is an attractive hypothesis to explain observations, there is at present no firm proof that a massive black hole exists in any of these galaxies or in any binary.

18

Pulsating stars

18.1   Period–density relation

In Volume 1 we saw that there is a group of stars which periodically change their size and luminosities. They are actually pulsating (the pulsars are not). When Leavitt (1912) studied such pulsating stars, also called Cepheids, in the Large Magellanic Cloud she discovered that the brighter the stars, the longer their periods, independently of their amplitude of pulsation. In Volume 1 we discussed briefly how this can be understood. The pulsation frequencies are eigenfrequencies of the stars. They are similar to the eigenfrequencies of a rope of length $2l$, which is fastened at both ends but free to oscillate in the center (see Fig. 18.1a). If you pull the rope periodically down in the center, first slowly and then more rapidly, you find that for a given frequency ν_0 a standing wave is generated in the rope. For this frequency you need to put in only a very small amount of energy, much less than for the other frequencies, for which running waves are generated which interfere with each other and are therefore damped rapidly. The frequency ν_0, which generates the standing wave, is an eigenfrequency of the rope. If you increase the *amplitude* of the wave you still find the same eigenfrequency ν_0. If you increase the *frequency* further you again find running waves until you reach another frequency ν_2, three times as large as ν_0, for which another standing wave is generated. This wave has two nodes and a wavelength which is a third of the wavelength for the eigenfrequency ν_0 (Fig. 18.1c). There also exists an eigenfrequency ν_1 which is twice as large as ν_0. For this frequency the wave has a node in the center (Fig. 18.1b). You would have to pull at a distance of $\frac{1}{2}l$ from the end of the rope in order to excite this wave. The only possible standing waves are those which satisfy the boundary conditions at both ends of the rope, namely that the amplitude has to be zero at these points. Because of this boundary condition a wave traveling along the rope is reflected at the end and travels back. If the incoming wave and the reflected wave always have the same phase at any given point x they enhance each other and we obtain a standing wave. If they do not have the same phase they interfere with each other and partially cancel each other. Such waves are strongly damped and new energy has to be fed in all the time.

The frequency ν_0 is called the fundamental eigenfrequency and $P_0 = 1/\nu_0$ the fundamental period. ν_1 is the first overtone or first harmonic frequency and ν_2 the second overtone or second harmonic, etc. The corresponding wavefunctions $\xi(x)$, shown for the rope in Fig. 18.1, describe the amplitude at any given point x, and are called the eigenfunctions for the different modes of pulsation.

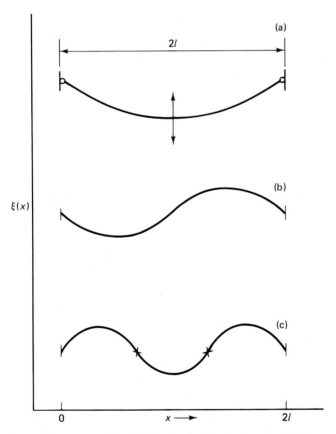

Fig. 18.1. (a) A rope of length $2l$ is fastened at both ends. It is free to oscillate in the center. It has a fundamental frequency ν_0 for a standing wave with nodes only on both ends. (b) The first overtone with frequency ν_1 for standing waves has a node in the center of the rope and the maximum amplitude at distance $d = \frac{1}{2}l$ from the walls. (c) The second overtone mode for standing waves has two nodes at distance $d = \frac{2}{3}l$ from the walls and maximum amplitude in the center and at distance $d = \frac{1}{3}l$ from the walls.

The period length is given by the travel time of the wave along the rope back and forth, namely $P = 4l/c$ where c is the phase speed of the traveling wave.

A star acts like the rope, in a way. While for the rope the stresses in the rope are the restoring forces, trying to bring it back to its equilibrium position and hence causing it to oscillate, the restoring forces for the star are the pressure forces and the gravitational forces. In the rope the stresses determine the phase speed of the wave; in the star the pressure forces determine the phase velocity of the wave, which is therefore the sound velocity. (Remember that a sound wave is also a traveling pressure disturbance.) For a spherically symmetric pulsation of a star the waves from all directions converge in the center and are reflected. Because of the convergence in the center the displacement amplitude in the center $\xi(0)$ must be zero, just as it is for the ends of the rope. At the surface of the star the amplitude can be large, just as for the center of the rope. The surface of the star corresponds to the point $x = l$ for the rope. The length of the period P is therefore $4R/\bar{c}_s$, where $\bar{c}_s$ is the average sound velocity. $c_s = \sqrt{\gamma(P_g/\rho)}$ for adiabatic sound waves and $\gamma = C_p/C_V = \frac{5}{3}$ for monatomic gas. As in Volume 1 we find

$$\frac{\overline{P_g}}{\bar{\rho}} \approx \frac{GM}{R} \quad \text{with} \quad \overline{P_g} \approx \bar{\rho}R\bar{g} \sim \bar{\rho}\frac{GM}{R} \tag{18.1}$$

With this we find for the period P

$$P \sim \frac{4R}{\sqrt{(\gamma(GM/R))}} \sim \frac{4}{\sqrt{(\gamma G)}} \frac{1}{\sqrt{\bar{\rho}}} \sqrt{\left(\frac{3}{4\pi}\right)} \tag{18.2}$$

or

$$P = \text{const.} \times \bar{\rho}^{-1/2} \quad \text{or} \quad P = Q(\bar{\rho}/\bar{\rho}_\odot)^{-1/2} \tag{18.3}$$

where Q is a constant (if $\gamma = \text{const.}$) and $\bar{\rho}_\odot$ is the average density of the sun, $\bar{\rho}_\odot = 1.4 \text{ g cm}^{-3}$. Equation (18.5) generally holds for adiabatic pulsations; see Section 18.4.

The larger the luminosity of the pulsating star, the larger the radius and the smaller the average density $\bar{\rho}$ and the longer the period. The larger the radius of the star, the longer the wave traveling time to the stellar center and back. From the theoretical point of view the observed period luminosity relation is a period density relation.

In stars the phase velocity of the waves, i.e. the speed of sound, is different at different places, because $c_s \propto \sqrt{T}$. The traveling time, as measured by the length of the period, therefore depends on the temperature stratification in the star. This means that the value of Q in equation (18.3) depends on the temperature stratification. The values of Q also depend on non-adiabatic effects which are different for different stars. For pulsating stars with different luminosities the values of Q vary between 0.01 and 0.09. For helium stars they may be quite different. As Q depends on the temperature and density stratifications, the observed values can be used to check our calculations.

In stars, temperature and density increase by orders of magnitude from the outside towards the inside. The eigenfunctions $\xi_i(x)$, which describe the wave amplitudes as a function of $x = r/R$, are therefore very different from those of a rope. In Fig. 18.2 we reproduce the eigenfunctions $\xi_i(x)$ as obtained for an (unrealistic) star of homogeneous density. More realistic models are the *polytropes* for which $P \propto \rho^{(n+1)/n}$. (For $n = \frac{3}{2}$ we find $P_g \propto \rho^{5/3}$ as for the adiabat with $\gamma = \frac{5}{3}$.) For different n, different degrees of mass concentration towards the center are found, as is seen in Table 18.1. In Fig. 18.3 we show the eigenfunctions $\xi_i(x)/x$ for the polytrope with $n = 3$ which comes close to a real star. For such a star with a large central density the amplitudes in the inner regions are extremely small (too much mass has to be moved). In all these models the nodes for the overtone pulsations occur at different values of r/R than in the case of the rope, because the

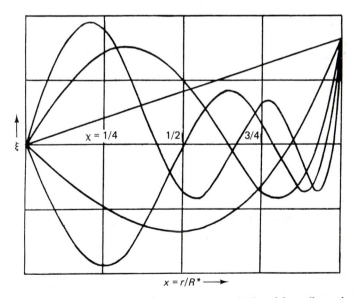

Fig. 18.2. The eigenfunctions $\xi_i(x)$ are shown as calculated for a (hypothetical) homogeneous density star. The amplitudes must be zero at the center. From Rosseland (1949).

Table 18.1. *Central density concentrations for different*
polytropes. From Rosseland (1949)

Polytrope index	0	2	3	4
$\rho_c/\bar{\rho}$	1	11.4	54.4	623

restoring forces in the star are different at different values of r/R, while for
the rope the stresses are the same everywhere.

The ratios of the frequencies v_1/v_0 and v_2/v_0 are also different from those
of the rope. They depend on the density stratifications in the stars. For
stars with homogeneous densities we calculate $v_1/v_0 = 3.56$ and
$v_2/v_1 = 1.56$, corresponding to the period ratios of $P_1/P_0 = 0.281$ and
$P_2/P_1 = 0.639$, while for the more realistic polytrope with $n = 3$ and
$\gamma = 1.54$ we find $P_1/P_0 = 0.687$ while $P_2/P_1 = 0.749$. The ratio P_1/P_0 is
increased by more than a factor of 2 as compared to the homogeneous
density model, but the value of P_2/P_1 is changed only slightly. It appears
that the ratio of P_1/P_0 is a very good measure for the central density
concentration. If we can measure this ratio it provides an excellent check
on the model pressure and temperature stratification.

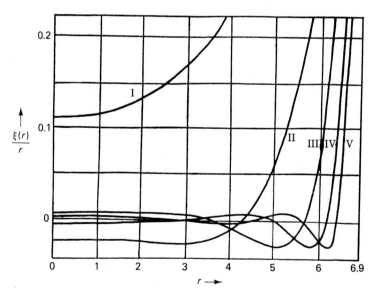

Fig. 18.3. The eigenfunctions $\xi_i(r)/r$ are shown for a star whose temperature and density
stratification follow a polytrope with index $n = 3$. For the larger central density, as
compared to the homogeneous star, the amplitudes in the center decrease strongly. The
index of the harmonics is given in roman numerals. I indicates the fundamental mode, II
the first harmonic etc. From Rosseland (1949).

For some RR Lyrae stars, which pulsate with periods of about 0.5 day, period ratios of $P_1/P_0 \sim 0.744$ are observed, confirming a high central density concentration. Some so-called beat Cepheids also show period ratios close to this value.

For realistic stellar models calculated using the Henyey method, we compute values of $P_1/P_0 = 0.74$ and $P_2/P_1 = 0.68$ for a 5 $M_\odot$ star.

18.2 Evolutionary state of Cepheids

In Volume 1 we saw that pulsating stars occur in the HR diagram along a narrow band which reaches from very luminous late G type ($T_{eff} \sim 5800$ K) supergiants to late A stars ($T_{eff} \sim 8000$ K) near the main sequence. The brightest stars in this band are the δ Cephei stars. The band is therefore called the Cepheid instability strip. In Fig. 15.2 we showed the position of this instability strip in the HR diagram together with the evolutionary tracks for stars with masses between 3 and 9 solar masses, according to calculations by Becker, Iben and Tuggle (1977). The δ Cephei stars occur in a region of the HR diagram for which the first evolution of stars from left to right is very fast, as we have seen in Chapters 14 and 15. There is little chance of seeing such stars on their first crossing through the instability strip. The massive stars, however, have another chance to cross the instability strip, during their 'blue loop' excursions. The evolution in the direction towards the blue, after starting helium burning in the center, is much slower than that from the blue to the red. On their second crossing of the instability strip the stars therefore stay much longer. During their blue loop phase they generally spend the longest time at the tip of the loops, as seen in Fig. 15.2. If for a given mass the tip of the blue loop falls into the instability strip we should find a relatively large number of Cepheids with this particular mass and period. For slightly lower masses the blue loops become shorter and no longer reach the instability strip, as seen in Fig. 15.1 for $M = 3 M_\odot$. For such low masses we should not find any Cepheids. Cepheids with the tips of their blue loops in the instability strip with masses just before the cut-off mass should have the shortest fundamental periods. These then should be relatively abundant, at least if we only look at Cepheids pulsating in the fundamental mode.

The blue loops extend further to the blue for stars with lower abundances of heavy elements. For these we expect to see lower mass Cepheids than for the high metal abundance population. In the Large and in the Small Magellanic Clouds with lower than solar metal abundances we therefore expect to see Cepheids of lower mass than in our galaxy.

Cepheids cross the instability strip only in a very advanced state of evolution, most of them having started helium burning in the center; this means they have a helium core which has contracted to very high densities and temperatures. The strong mass concentration in the center of the Cepheids is very important for the identification of the excitation mechanism for the pulsations.

18.3   Analysis of pendulum oscillations

In order to understand better why the Cepheids start pulsating let us look at the simple example of an oscillating pendulum. The equilibrium position of the pendulum is the lowest one, where the gravitational force is balanced by the tension in the string. If we lift the pendulum up it will fall back towards its equilibrium position. Down to this point it is accelerated by gravity. At the lowest point it is no longer accelerated but because it has a high velocity it overshoots the equilibrium position due to its inertia and moves upwards. During its climb gravity accelerates it downward and finally makes it stop at a height equal to its original height on the other side (see Fig. 18.4). Gravity then reverses the velocity of the pendulum and it falls down again towards its equilibrium position where it overshoots, etc. The process would continue forever if there were no damping due to friction in the hanging point and due to the resistance of the air.

The pendulum, of course, will never start by itself to swing; it needs a starting push. When stars evolve into the instability strip they do not

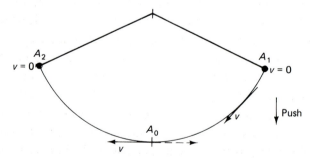

Fig. 18.4. If a pendulum is lifted to position A_1 it falls down to its equilibrium position A_0. It arrives there with a velocity v and due to its inertia overshoots to position A_2. At this point is has used all its kinetic energy to work against the gravitational pull downwards. It falls back and overshoots again, etc. The amplitude can be increased by a small downward push while the pendulum is falling. A small downward push while it is climbing decreases the amplitude.

pulsate while they are on the cool side of the strip. When they reach the red boundary of the strip they suddenly start pulsating all by themselves. How can this happen? Let us look again at how we could excite the pendulum to swing. Suppose we give it a very small push to begin with and then each time when it starts to fall we give it a little push downward. In this way the amplitude will slowly increase. We have to be very careful, however, to give the push at exactly the right phase of the oscillation. If we push downwards while the pendulum is still moving upwards, the amplitude will decrease and we will damp the oscillation. If we give completely symmetric pushes, for instance downwards while the pendulum is moving upwards and the same downward push while it is moving downward, nothing will change. In other words, if we increase the restoring force all the time we cannot excite the pendulum. **An increase in gravity would not increase the amplitude.** We need to push harder downwards when it is moving downward than we push down when the pendulum is moving upwards.

The stars in the Cepheid strip manage to give themselves a push at the right phase to increase their amplitude. There are several ways in which the star can do this. We discuss here the one which seems to be the most important. It is called the κ mechanism because it is due to changes in the absorption coefficient. In order to understand how it works we first have to look at adiabatic pulsations, where no energy exchange takes place between different layers in the star.

18.4   Adiabatic pulsations

Once a star is brought out of its equilibrium configuration the interplay of pressure and gravitational forces causes periodic inward and outward motions of the star similar to that of the pendulum, except that for the pendulum it is always gravity which pulls it back to its equilibrium position. For the star the pressure forces push it outwards when it is too small and the pressure forces are too large to be balanced by the gravitational forces. The gravitational forces pull it back when it is too large and the pressure forces are too small to balance gravity. The star will overshoot the equilibrium position like the pendulum. If there is no energy exchange between the different layers during the pulsation, we would have adiabatic contractions and expansions which could go on forever. The temperature would be highest when the star has its smallest radius and lowest when it has its largest radius (see Fig. 18.5). In Volume 1 (see Fig. 18.6) we saw that the observed light and radial velocity variations of a

Cepheid do not correspond to such adiabatic pulsations. The highest temperature is observed when the star is expanding and has nearly reached its equilibrium radius. The lowest temperature occurs when the star is moving inwards and has again nearly reached the equilibrium value of the radius. Clearly the pulsations are not adiabatic. We know that there must be some damping, because the radiative energy losses are proportional to T^4. The star loses more energy during high temperature phases when the excess gas pressure expands it. This energy loss decreases the temperature and thereby reduces the pressure forces which push the star out again. The restoring force is thus reduced and the star does not expand to the same large radius it did one period earlier. The pulsation is damped unless there is also an extra push outwards at each period. This is the case for most stars, and is the reason why most stars are not pulsating.

If the pulsations are to be excited, heat energy has to be added mainly at phases shortly after maximum compression so that the restoring force, the pressure, is increased during the outward acceleration.

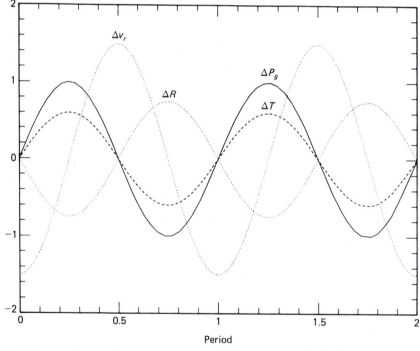

Fig. 18.5. The velocity, radius, pressure and temperature variations for adiabatic pulsations are shown schematically. Pressure and temperature are highest for the smallest radius. For excess pressure, the layers are accelerated outwards. For decreased pressure, gravity pulls inwards. Arbitrary units.

18.5   Excitation of pulsations by the *κ* mechanism

In Chapter 7 we discussed that we expect a rather sharp dividing line in the HR diagram between stars with a hydrogen convection zone and those without. Calculations show that this theoretical dividing line agrees with the observed position of the Cepheid instability strip.

As was discovered by Zhevakin (1959) the temperature and pressure dependence of the absorption coefficient can for these stars feed energy into the pulsations. The energy is taken out of the radiative flux. The mechanism was explained by Baker and Kippenhahn (1962). In Fig. 18.7 their picture of the κ 'mountain' is reproduced. This is a three-dimensional display of the dependence of the absorption coefficient on temperature and pressure. During adiabatic contraction, temperature and pressure both increase; κ may increase or decrease depending on whether we go

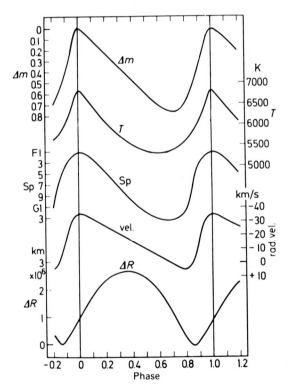

Fig. 18.6. The observed variations of the apparent visual magnitudes, the effective temperatures, the spectral types and the radial velocities for δ Cephei are shown as a function of phase. At the bottom the changes in radius are shown as obtained by integrating over the pulsational velocities. The radius is nearly the same for maximum and minimum brightness. (Adapted from Becker, 1950. See also Volume 1.)

uphill or downhill on this mountain. From our discussion in Chapter 3 we know that the radiative flux F_r is given by

$$F_r = \frac{4}{3} \frac{1}{\kappa} \frac{dB}{dz} \quad \text{where} \quad dz = -dr \qquad (18.4)$$

If κ increases during compression, F_r decreases. If a certain amount of flux comes from the deeper layers it cannot propagate easily through the layer with an increased κ. The energy flux gets stuck in the high κ region. The energy accumulates and the region increases its temperature. As soon as κ starts to increase the layer heats up. It is important to note that heating continues as long as κ is larger than in the equilibrium configuration. A higher temperature means a higher pressure, or a pressure excess above the pressure which would be present for adiabatic contraction (see Fig. 18.8). Since heating continues throughout the compression phase, the highest excess temperature and pressure is reached when the star reaches the equilibrium radius. Then the star expands and during expansion temperature and pressure decrease and κ decreases. This means the radiative flux F_r now increases above its equilibrium value, excess energy is taken out and the layer cools finally below its equilibrium value. This reduces temperature and gas pressure during the expansion phases. The

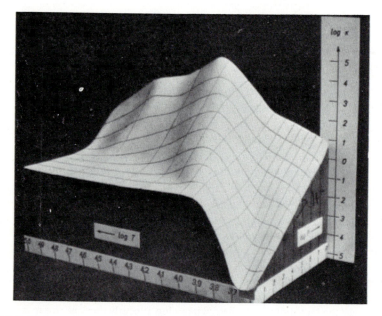

Fig. 18.7. The κ mountain as constructed by Baker and Kippenhahn (1962).

pressure reduction continues throughout the expansion time. The gravitational pull-back is therefore more efficient due to the reduction of the opposing force. In Fig. 18.8(a) the sinusoidal pressure variation expected for adiabatic pulsation is plotted schematically. Fig. 18.8(c) shows a schematic plot of the excess temperature $\Delta\Delta T$ leading to an excess pressure $\Delta\Delta P_g$ generated by the heating or cooling due to the changing κ in a layer where κ increases during adiabatic contraction and decreases during adiabatic expansion. It is important to realize that $\Delta\Delta P_g$ has a

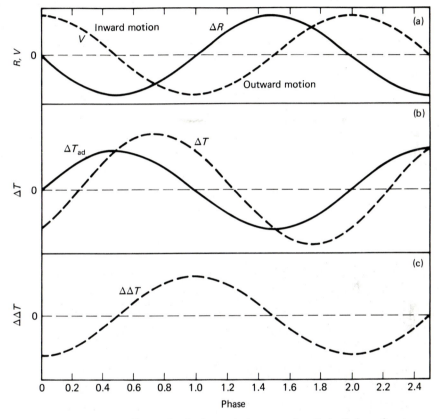

Fig. 18.8. (a) Adiabatic radius and velocity variations as a function of phase for a Cepheid (arbitrary scale). Negative velocities mean outward motion, positive velocities falling motion. (b) The adiabatic temperature variations ΔT_{ad} (solid line). Maximum ΔT_{ad} occurs for minimum radius, and vice versa. In those layers where κ is increasing with increasing T and P_g, excess heating occurs during phases of positive ΔT and ΔP_g. (Excess cooling occurs for negative ΔT and ΔP_g.) (c) The excess temperature increase, $\Delta\Delta T$ is shown schematically. It increases throughout the phase of positive ΔT and ΔP_g. Only after expansion has proceeded beyond the equilibrium radius does $\Delta\Delta T$ actually decrease. In (b) the final $\Delta T = \Delta T_{ad} + \Delta\Delta T$ is shown schematically as a function of time (dashed line). $\Delta P_g \propto \Delta T$ is larger during expansion phases than during contraction phases, leading to excitation of the pulsation.

phase shift of $\sim\frac{1}{2}\pi$ in comparison with the adiabatic pressure changes. Fig. 18.8(b) shows a schematic plot of the final temperature (and pressure) variations as a function of phase. The final gas pressure change has a phase shift with respect to the adiabatic case as seen by comparison with Fig. 18.5. During the early expansion phase after the star has just had its smallest radius, the excess gas pressure gives an excess push outwards while the gas is already moving out. The excess pressure is zero or very small when the star is still moving inwards. This corresponds to giving the pendulum an extra push just after it has reached its maximum height and is moving down already, exactly as we need it for the excitation of the oscillation. During the phases of increased radius the pressure *deficit* increases and reaches its maximum when the star is already shrinking. This corresponds to giving the pendulum another push downwards when it is falling down already. The restoring force, in this case gravity, becomes more effective when the star is shrinking than when it is expanding. We find exactly the phase relations we need for the enhancement or excitation of the pulsation. This, of course, only works in the regions where κ increases during contraction. If κ decreases during contraction the opposite will be true and the pulsation is damped. In Fig. 18.7 we see that next to a region where κ increases during contraction there is also always a region where κ decreases during contraction, i.e. a damping region. Some layers contribute to excitation, some to damping.

We know that for temperatures above $\sim$6000 K κ increases because of the beginning hydrogen ionization and the excitation of higher energy levels in the hydrogen atom. We also know that this increase in κ leads to convective instability in the hydrogen ionization zone. Another increase in κ for adiabatic contraction occurs in a temperature range around 40 000 K where the Lyman continuum absorption of hydrogen becomes very important and where the helium ground level absorption contributes to the opacity. This is also the region where He^+ ionizes to He^{2+}. Both regions are convectively unstable. Between these two κ ridges we find a valley which contributes to the damping of the pulsations. The question then is which is more efficient: the excitation due to the increasing κ during contraction or the damping due to the regions of decreasing κ during contraction? The answer depends strongly on the amplitude of the pulsation in the different layers, which is determined by the eigenfunctions $\xi(r)$. If the amplitudes are large in the excitation regions and much smaller in the damping regions, then we find a net excitation. If the amplitudes are larger in the damping regions then the star is stable at least for that particular mode of pulsation. The amplitude of the fundamental mode may be large in an excitation region and small in a damping region. If so, the star is unstable for pulsations in the fundamental mode. For the same star the amplitude for the first harmonic mode may be small in the excitation region and larger in the damping region. The same star would then be stable for pulsations in the first harmonic mode. The opposite may be true for another star. It is indeed found that the instability strips for the different modes do not completely overlap, as seen in Fig. 18.9 where we show the blue boundaries for the instability strip for a star with $M = 0.6\,M_\odot$. The boundary lines for the fundamental mode and for the first harmonic mode do not agree. For the lowest luminosities the bluest variables can only pulsate in the first harmonic mode, while for the more luminous stars the bluest variables are unstable only to the fundamental mode.

18.6   Excitation by nuclear energy generation?

In the early discussions of pulsational instability nuclear reactions were considered to be a likely excitation mechanism. During contraction temperature and pressure in the burning regions are higher leading to an

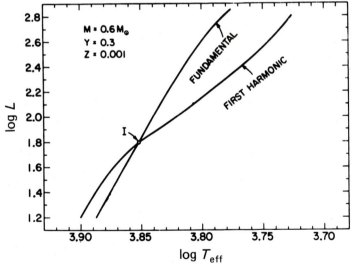

Fig. 18.9. In the luminosity, T_{eff} diagram the blue edges of the instability strip for the fundamental mode and for the first harmonic mode are shown for stars with $M = 0.6\,M_\odot$ and different luminosities. For higher luminosities the instability in the fundamental mode extends to higher T_{eff} than that for the first harmonic mode. For lower luminosities the inverse is true. On the horizontal branch the bluest RR Lyrae stars, Bailey's type c (see Volume 1), pulsate in the first harmonic mode. The cooler, type a RR Lyrae stars pulsate in the fundamental mode. Adapted from Iben (1971).

increased energy generation and to heating of this region, which leads to an increase in the gas pressure which also has its maximum during the early expansion phases. Heating continues until the star has reached the equilibrium radius. For larger radii the energy generation decreases and the interior cools. The effect is very similar to that of an increase in κ in the outer regions. The phase relations are also the same. This indeed makes the nuclear burning region an excitation region. In this context it is, however, important that the Cepheids are in an advanced state of evolution with a very high density concentration in the interior. The pulsation amplitudes, the eigenfunctions $\xi(r)$, are therefore extremely small in the region of nuclear energy generation; see Fig. 18.3 for this effect. Because of this extremely small amplitude the excitation by the nuclear energy generation is so minute that it could never overcome the damping in the surface regions where we find large amplitudes. If the nuclear burning region were further out in the star where the amplitudes are larger this excitation mechanism would be more efficient. Also were the temperature dependence of the energy generation much stronger this excitation mechanism could be much more efficient. J. P. Cox (1956) calculated that nuclear burning excitation could overcome the damping if the burning region was further out than about 50 per cent of the stellar radius or if the temperature dependence of the energy generation was proportional to T^{90}. The latter can never be the case. The former could possibly happen for very large stages of stellar evolution when a star has lost most of its envelope.

18.7   Limits of pulsation amplitudes

We now ask how the pulsations start and what limits the amplitudes. What stops the amplitude from growing indefinitely? The latter question is easy to answer if we look at the κ mountain in Fig. 18.7. If during contraction the change in temperature and pressure becomes so large that we climb over the top of the mountain ridge then a further increase in amplitude causes a decrease in κ and therefore damping. The pulsational amplitude is limited by the height of the ridges in the κ mountain. Other effects, such as the increasing energy loss at the surface for very large temperature variations, also limit the amplitude.

In order to start the pulsation we only need a very small accidental increase in pressure at some point. This will then increase in time and lead to pulsations. The original perturbation needs to be only infinitesimally small. It will inevitably grow.

Such small disturbances may occur easily in a convectively unstable region even if the velocities and the convective energy transport are very small.

18.8 The edges of the instability strip

18.8.1 *The blue edge*

Regions of larger optical depth are more efficient in generating excitation or damping. For low optical depth the radiation can escape more easily and any temperature differences are most easily smoothed out. If the excitation region moves closer to the surface for stars with increasing effective temperatures the pulsational instability ceases. This determines the blue edge of the instability strip. Since the excitation regions are at different depths for the fundamental mode and for the first harmonic the blue edges for the two modes are at different positions in the HR diagram. Stars close to the blue edge may be unstable only to the fundamental mode or only to the first overtone depending on their L and T_{eff} (see Fig. 18.9).

18.8.2 *The red edge*

We saw that the excitation by the κ mechanism is due to the increase in κ in the convectively unstable zone which leads to a trapping of the radiative flux, which in turn leads to an increase in the temperature. This can only work if there is a radiative flux which can be trapped by an increase in κ. If most of the energy is transported by convection then an increase in κ cannot trap this energy. The remaining small amount of radiative flux cannot contribute much to excess heating. If we have efficient convective energy transport the excitation mechanism cannot work. Pulsations can only be excited in stars in which the convective energy transport is inefficient in the regions of increasing κ. This occurs just at the boundary line for the onset of efficient convection, exactly where we observe the pulsating stars in the HR diagram, namely in the Cepheid instability strip. If the Cepheids are close to the red boundary line they may perhaps develop efficient convection for a small fraction of their pulsational cycle. Whether this actually happens is at the moment not clear.

It might be mentioned that several of the discussions in this chapter have to be modified when discussing pulsations of massive, hot stars, for which radiation pressure becomes very important.

19

The Cepheid mass problem

19.1   Importance of Cepheid mass determinations

We know that Cepheids must be in an advanced state of evolution because the blue loops are the only way they can stay in the instability strip for any length of time. If we can determine mass and luminosity for a Cepheid we can check whether its luminosity agrees with what we expect without overshoot or additional mixing. A larger L might indicate additional mixing (see Fig. 15.3). In fact we could calibrate the amount of mixing for the Cepheid progenitor on the main sequence by determining mass and luminosity for a given Cepheid. Of course, we also have to know the chemical abundances and the correct κ. For a given L the derived masses of the Cepheid may differ by 50 per cent if for instance the assumed helium abundance is changed by a factor of 2.

We can also check the consistency of the stellar evolution and pulsation theories by determining masses of Cepheids in different ways, making use of either evolution or pulsation theory or of different aspects of those theories. If the theories are correct we should, of course, find the same mass, no matter how we determine it.

19.2   The period–luminosity relation

A number of Cepheids are found in galactic clusters. Their periods can be measured and their distances can be determined, for instance, by main sequence fitting or equivalent methods. We can thus find their absolute magnitudes averaged over one period. The first extensive study of distances for clusters with Cepheids was done by Sandage and Tammann (1968), and a more recent one was done by Schmidt (1984). A period–luminosity relation can then be derived. Sandage and Tammann determined cluster distances by fitting the cluster main sequence with the main sequence of the Hyades cluster for which they adopted

238

$m_{V0} - M_V = 3.0$. The recently determined value is $m_{V0} - M_V = 3.30$. Using this new distance modulus Sandage and Tammann (1968) find a period–luminosity relation

$$M_V = -3.42 \log P + 2.52(\bar{B}_0 - \bar{V}_0) - 2.16 \qquad (19.1)$$

The more recent determination of the period–luminosity relation by Schmidt (1984) yields

$$M_V = -3.8 \log P + 2.70(\bar{B}_0 - \bar{V}_0) - 2.21 \qquad (19.2)$$

which gives somewhat fainter magnitudes for a given period and therefore smaller distances to the Cepheids. $\bar{B}_0$ and $\bar{V}_0$ are the average blue and visual magnitudes of the Cepheid, corrected for interstellar absorption.

19.3 Evolutionary masses

With the period–luminosity relations (19.1) or (19.2) we can also derive distances and absolute magnitudes for single Cepheids for which periods and colors have been measured. Determining T_{eff} from the average $B - V$, the luminosity from M_V and the bolometric corrections BC the position of the Cepheid in the T_{eff} luminosity diagram can be fitted on an evolutionary track for a given mass which should give us the mass of the Cepheid. These masses are called evolutionary masses. With $L \propto M^{4.5}$ an error in the distance modulus by $\Delta M_v = 0.3$ or $\Delta \log L = 0.12$ gives an error in the mass by $\Delta \log M = 0.026$ or by 6 per cent. The derived evolutionary masses are rather insensitive to small errors in the distance determination. Fig. 16.1 shows that for a given luminosity smaller evolutionary masses are found for higher helium abundances and smaller abundances of heavy elements. For instance for the Cepheid S Muscae which has a luminosity of $\log L \sim 3.5$, a mass of $M/M_\odot = 5.5$ is derived if $Y = 0.36$ and $Z = 0.02$ while a mass of $M/M_\odot = 7.5$ is found for $Y = 0.20$ and the same Z. For $Y = 0.28$, the currently adopted value, $M/M_\odot \sim 7.5$ would be found for $Z = 0.03$ and $M/M_\odot = 5.5$ for $Z = 0.01$.

19.4 Pulsational masses

We saw in Chapter 18 that the observed period–luminosity relation can theoretically be traced back to a period–density relation

$$P = Q\left(\frac{1}{\bar{\rho}/\bar{\rho}_\odot}\right)^{1/2} \qquad (19.3)$$

The 'constants' Q are dependent on the mass and on the temperature and density stratifications in the star. For a given stellar model Q can be calculated. From the observed pulsation period P of a given Cepheid and from the calculated Q the average density

$$\bar{\rho} = \frac{M}{\frac{4\pi}{3}R^3}$$

can be determined using equation (19.3). Knowing T_{eff} of the Cepheid from its B − V color and $L = 4\pi R^2 \sigma T_{\text{eff}}^4$ (from M_V and BC) the radius can be found and $M \propto R^3 \bar{\rho}$ is obtained. If the observed luminosities (i.e., the distance determinations), T_{eff} (which has only minor influence) and our theoretical models are correct, we should find the same values for these masses as those derived for the evolutionary masses. Unfortunately the pulsational masses M_p depend sensitively on the adopted distances d of the stars because $R^2 \propto L$ and therefore $M_p \propto L^{3/2}$. With $L \propto D^2$ we find $M_p \propto d^3$. For distances as determined according to equation (19.1) the derived pulsational masses nearly agree with the evolutionary masses M_{ev}. For the smaller distances determined from equation (19.2) pulsational masses smaller by about 40 per cent are found. An error of $\Delta M_V = 0.3$ leads to $\Delta \log M = 0.18$, or 50 per cent error in the derived mass.

19.5   Baade–Wesselink masses

In order to determine reliable masses of Cepheids we need methods which are independent of the still uncertain distance determinations. The Baade–Wesselink method is one we can use. The radii of the stars can be determined from the observed radial velocities v_r from which the expansion velocities v_{exp} can be derived. We find for the difference in Cepheid radii at phase ϕ_1 and ϕ_2 corresponding to time t_1 and t_2

$$\Delta R = R(\phi_2) - R(\phi_1) = \int_{t_1}^{t_2} v_{\text{exp}}(t)\, dt \qquad (19.4)$$

We can measure radial velocities, v_r, from spectral lines. From Fig. 19.1 we see that the measured radial velocity is the average of the projected lines of sight component of the expansion velocity averaged over the visible surface of the star. The factor v_{exp}/v_r is uncertain (see Figs. 19.1 and 19.2) but probably close to $v_{\text{exp}}/v_r \sim 1.3 \pm 0.1$ for the Cepheids discussed here. In order to determine the radius itself the variations in bolometric magnitudes are used:

$$m_{bol}(\phi_1) - m_{bol}(\phi_2) = 2.5 \log \frac{L(\phi_2)}{L(\phi_1)} = 2.5 \left[\log \frac{T_{eff}^4(\phi_2)}{T_{eff}^4(\phi_1)} + \log \frac{R^2(\phi_2)}{R^2(\phi_1)} \right]$$

(19.5)

The T_{eff} for the two phases are determined from the measured B − V colors or from infrared colors. With T_{eff} known the ratio of the radii $X = R(\phi_2)/R(\phi_1)$ is obtained and $R(\phi_1) = \Delta R/(X - 1)$ can be calculated. With $R(\phi_1)$ determined for many phases, the average radius can be calculated and used in equation (19.3) to estimate the mass from the pulsation period. These masses are called the Baade–Wesselink masses,

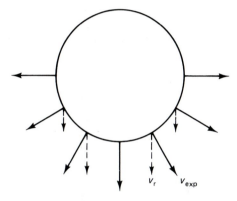

Fig. 19.1. The absorption line being formed at the surface of the star shows the average line of sight velocity component v_r of the material expanding with the velocity v_{exp}. The v_r is considerably smaller than the expansion velocity. For a lower limb brightness the average v_r is somewhat larger, and the ratio v_{exp}/v_r becomes smaller.

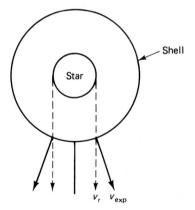

Fig. 19.2. If line absorption takes place in an expanding shell around the star only the material in front of the star can contribute to the absorption. For this material the line of sight velocity component v_r is nearly equal to the expansion velocity v_{exp}.

M_{BW}. The method is very nice in principle. Unfortunately at present different radii are obtained depending on whether B − V colors or infrared colors are used. Using B − V colors the M_{BW} come out to be much smaller than M_{ev}. For infrared colors the M_{BW} agree better with M_{ev}. There are apparently still problems with the T_{eff} color calibrations and with the determination of the correct ratio v_{exp}/v_r to be used.

19.6   Bump masses and beat masses

19.6.1   Bump masses

There is another way to determine Cepheid masses independently of the distance determinations. This method makes use of the shape of the observed magnitude and velocity curves. For Cepheids with periods around 10 days the light and velocity curves show bumps (see Fig. 19.3). The phases of these bumps shift systematically with increasing period. Such bumps occur also in calculated light and velocity curves. Theoretically the phase for which this bump occurs depends on the mass of the Cepheid. The observed phase of the bump can therefore be used to determine the so-called bump masses, M_b. These turn out to be about half

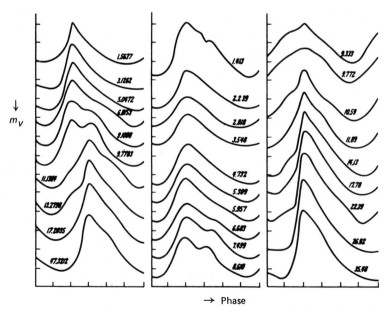

Fig. 19.3. For Cepheids with periods around 10 days the light (and velocity) curves show bumps. The phase of the bump shifts systematically for increasing periods. The Cepheid periods in days are noted on the light curves. From Ledoux and Walraven (1958).

as large as the standard evolutionary masses M_{ev} but agree fairly well with the pulsational masses for the *smaller* distances.

19.6.2 Beat masses

Some short period Cepheids show the so-called beat phenomenon. Basically they pulsate with one period but the amplitude of the pulsation changes periodically with a longer period. Such beat phenomena are observed when the pulsation results from the superposition of two periods: If, for instance, the ratio of the periods is 3:4 then after 3 of the longer periods and 4 of the shorter ones the two pulsations are in phase and the amplitudes add up, so the pulsation amplitude is large. Halfway in between, the pulsations in the two periods are in antiphase, the amplitudes are subtracted and the final pulsation amplitude is small. We saw in Chapter 18 that the ratio of the period length for the first harmonic and the fundamental period depend on the degree of mass concentration in the center, which in turn depends on the mass. For smaller masses the density concentration is somewhat larger. The period ratios can therefore also be used to determine the masses of these Cepheids. Very small masses, around one to two solar masses, are presently found in this way for the beat Cepheids. For such small masses the theoretical blue loops never enter the Cepheid instability strip, which does not agree with theoretical models. The real stars may have a stronger mass concentration than we think. Wrong values for the absorption coefficients could give wrong results or maybe these stars pulsate in the first and second harmonic modes and not in the fundamental and first overtone mode as assumed so far.

19.7   Dynamical masses

So-called dynamical masses of the second component of a binary system can be determined if the mass of the first component is known from other observations. For instance if star 1 is a main sequence star of spectral type A0 V we know that its mass M_1 must be around $M_1 = 2.5\ M_\odot$. In a binary system the mass ratio of the stellar masses M_2 to M_1 can be obtained from the ratio of the orbital velocities v_2 and v_1: at any given time we must have $v_2/v_1 = M_1/M_2$. If we can measure v_2/v_1 the dynamical mass M_2 can be obtained.

In order for the stars to have radial velocities large enough to be measured accurately the stars must be fairly close binaries with periods of a few years at most, which means their orbital radii can only be a few astronomical units. For Cepheid binaries they cannot then be visual binaries because the Cepheids are too far away for binaries with orbits of a few years to be resolved. We also must be able to measure radial velocities for both binary components of a spectroscopic binary system. For Cepheids this condition used to be difficult to meet, because they are bright supergiants and any main sequence companions are much fainter in the visual spectral region. The chances of finding a supergiant companion are extremely small. Even giant companions are too faint in the optical region. The availability of satellites, permitting observations in the ultraviolet, has changed this situation dramatically. The Cepheids have rather low temperatures ($\sim$5500 K) and emit very little light at short wavelengths. If a Cepheid has, for instance, a main sequence companion of spectral type B9 then the B9 star fluxes are larger than those for the Cepheid for wavelengths shorter than 2500 Å. For longer wavelengths the spectrum of the Cepheid dominates. The so-called 'long' wavelength camera of the International Ultraviolet Explorer Satellite (IUE) is sensitive for wavelengths 2000 Å $< \lambda <$ 3200 Å. On these spectra the radial velocities of both stars can be measured. For the Cepheid we have of course to correct for the pulsational velocities in order to determine the orbital radial velocities. Current results for several systems have been plagued by several observational problems and are still somewhat uncertain but the indications are that the dynamical masses come out to be lower than standard evolutionary masses and are close to the bump masses and the pulsation masses for the smaller distances. If confirmed it means that future stellar evolution calculations will have to be corrected to give smaller evolutionary masses for Cepheids, probably by including additional mixing for massive main sequence stars, by adopting larger helium or smaller heavy element abundances, or by re-evaluating the calculations of κ.

20

Star formation

20.1 Introduction

The fact that we see massive, luminous stars which cannot be older than about 10^6 years tells us that stars must have been formed within the last million years. In association with these luminous young stars we often see some peculiar stars with emission lines, called the T Tauri stars (see Volume 1). These can therefore be assumed to be young stars also. They have lower luminosities and are more red than the massive O and B stars but are considerably more luminous than main sequence stars of the same color. Because of their lower luminosities they must have lower masses than the O and B stars. For the lower mass stars the contraction times are longer, as we have seen in Chapter 2, because these stars cannot radiate away the surplus gravitational energy as fast as the more luminous, massive stars. If these lower mass T Tauri stars were formed at the same time as their more massive associates they have not had enough time to contract to the main sequence during the main sequence lifetime of the massive stars. Lower mass stars must therefore still be in the contraction phase. It is then reasonable to assume that these T Tauri stars are young stars still in the contraction phase.

Both kinds of stars, the massive O main sequence stars and the less massive young T Tauri stars, appear in association with large dust complexes, i.e. regions of high density where many interstellar molecules are formed. It thus appears that new stars may be born in regions of high density interstellar material.

In such regions we now see hundreds of young stars. In the somewhat older galactic star clusters in the galactic plane, we see comparable numbers of stars (see Fig. 1.3). On the other hand, the very old so-called globular star clusters have up to a million stars (see Fig. 1.5), which are all metal poor and apparently very old. Why did such very massive star clusters form in the early stages of our galaxy but no longer appear to do so? The observations of star clusters in the Magellanic Clouds, the nearest extragalactic systems, may give us some clues. In these external galaxies we see even now populous clusters with many massive young stars, like the SMC cluster NGC 330 (see Fig. 17.6). The Magellanic Cloud galaxies are much smaller than our own galaxy, and appear to have much less dust and a somewhat lower abundance of heavy elements. These conditions may be reminiscent of those of our own galaxy at an early stage and are apparently more favorable to the formation of very massive star clusters.

In the following discussion of star formation we shall see under which conditions we think that new stars and star clusters can be born, and we shall see whether we can understand why massive star clusters may be born no later than in the early stages of a galaxy.

20.2   Jeans' criterion for gravitational instability

Observations show that new stars can be formed in an environment of dense interstellar clouds. We will study here the circumstances under which a dense interstellar cloud can become gravitationally unstable to contraction; in other words we ask under which conditions the gravitational force, trying to contract the cloud, can overcome the gas pressure forces that try to expand the cloud. This question was first studied by Jeans, who considered a homogeneous gas volume in which he introduced a sinusoidal density perturbation; he checked under which conditions such a perturbation would grow, leading to a collapse of the higher density regions. Our approach here will be less accurate and less mathematical but, we think, physically more transparent. We consider an isothermal gas cloud in low density surroundings and focus our attention on a small volume of gas at its surface. We also assume a spherical cloud, which is a very great simplification, but the result obtained will be correct by order of magnitude. A volume element of 1 cm^3 with mass ρ at the surface of this sphere experiences two forces. One is the gravitational force $F_g = \rho(GM/R^2)$, if M is the mass of the cloud and R its radius. Due to any pressure gradient this volume element also experiences a force, $F_p = dP/dr$ (see Chapter 2). The equation of motion for this volume element is therefore

$$\rho \frac{d^2r}{dt^2} = -\rho \frac{GM_r}{r^2} - \frac{dP_g}{dr} \qquad (20.1)$$

where t stands for time.

For an element at the surface of the cloud $M_r = M$ and $r = R$. Dividing (20.1) by ρ gives

$$\frac{d^2r}{dt^2} = -\frac{GM}{R^2} - \frac{1}{\rho}\frac{dP_g}{dr} \tag{20.2}$$

where $P_g = (\rho T/\mu)R_g$. The equation of motion can then be written as

$$\frac{d^2r}{dt^2} = -\frac{GM}{R^2} - \frac{R_g T}{\mu}\frac{1}{\rho}\frac{d\rho}{dr} \tag{20.3}$$

With $d\rho/dr < 0$ the pressure term becomes greater than 0 which means it gives a positive, outward acceleration while the first term opposes this force. We have $d^2r/dt^2 < 0$, i.e., we find contraction, if

$$\frac{GM_r}{R^2} > \frac{R_g T}{\mu}\frac{1}{\rho}\left|\frac{d\rho}{dr}\right| \tag{20.4}$$

As a very crude approximation we now replace $|d\rho/dr|$ by ρ/R. We also make use of the relation

$$R = \left(\frac{M}{\rho}\frac{3}{4\pi}\right)^{1/3} \tag{20.5}$$

and obtain

$$\frac{GM}{R} = M^{2/3}\rho^{1/3}\left(\frac{4\pi}{3}\right)^{1/3}G \tag{20.6}$$

Making use of this and of equation (20.4) we find instability for contraction (i.e., inward acceleration) if

$$M \geqslant M_{\text{crit}} = \left(\frac{R_g T}{\mu G}\right)^{3/2}\left(\frac{3}{4\pi}\right)^{1/2}\frac{1}{\sqrt{\rho}} \approx \frac{T^{3/2}}{\rho^{1/2}}\frac{1}{2}\left(\frac{R_g}{\mu G}\right)^{3/2} \tag{20.7}$$

The mathematical instability analysis gives a factor $\pi^{3/2}$ instead of our factor $(3/4\pi)^{1/2}$. The main reason for the difference is that we have considered an isolated cloud with a pressure gradient $dP_g/dr = P_g/R$, while the instability analysis considers a sinusoidal density perturbation and in a homogeneous surrounding medium which also exerts forces. In the following numerical examples we use the factor $\pi^{3/2}$.

For small densities the gravitational forces are small and only large masses become gravitationally unstable. The limiting mass is very large and increases with increasing temperatures, because the opposing pressure forces increase with increasing temperature. For normal interstellar densities of the order of one particle per cm^3 (i.e., for $\rho \sim 10^{-24}$ g cm^{-3} and $\mu = 1$) we find if $T = 10$ K that a gas cloud will contract if $M \geqslant 3 \times 10^3 \, M_\odot$, while for a cloud with a temperature of 100 K, which is the average temperature of the cool interstellar medium in our galaxy, we find $M \geqslant 10^5 \, M_\odot$, close to the masses we find in globular clusters. For lower densities, as we find them in the galactic halo and as may be expected before the galaxy collapsed into a disk, we find that still larger masses are needed for gravitational instability.

On the other hand, in dense dust clouds or dense molecular clouds we have about 10^2 to 10^4 particles per cm^3 and the temperature may be as low as 10 K. For 10^2 particles per cm^3 and $\mu = 1$ we then find that $M_{crit} \geqslant 3 \times 10^2 \, M_\odot$. For larger μ, due to the formation of molecules and dust grains, even smaller critical masses are obtained. We may then even find masses of the order of the largest stellar masses to become gravitationally unstable. For a cloud of one solar mass to become gravitationally unstable we would need, however, a density of 10^4 particles per cm^3 if $\mu = 2$. Such high densities are not observed in interstellar clouds. We will discuss below how such low mass stars can possibly be formed. First we will follow what happens when a large, gravitationally unstable gas cloud of low density contracts.

20.3   Adiabatic contraction or expansion of a homogeneous cloud

In Fig. 20.1 we show in a temperature density diagram schematically the equilibrium line where for a given mass gravitational and pressure forces are in equilibrium, i.e., where

$$\frac{GM}{r^2} = \left| \frac{1}{\rho} \frac{dP}{dr} \right|$$

According to equation (20.7) this requires $T^{3/2} \propto \rho^{1/2}$. For points above this line, i.e., for higher temperatures, the pressure forces are larger than the gravitational forces, the gas cloud expands and cools adiabatically if there is no energy exchange with the surroundings. For temperatures and densities below the equilibrium line the gravitational forces are larger than the pressure forces, the cloud contracts adiabatically if there is no energy exchange with the surroundings, and the temperature increases. All the gas clouds of the given mass whose T and ρ combinations fall below the equilibrium line remain unstable against gravitational contraction as long as they stay below the equilibrium line. We now follow the track of a contracting cloud in this T, ρ diagram. In Fig. 20.1 we have plotted adiabats for a monatomic gas with the ratio of the specific heats $\gamma = \frac{5}{3}$. For such an adiabat we have $T \propto \rho^{2/3}$. If a cloud with T and ρ originally corresponding to point A in Fig. 20.1 would contract adiabatically, it would follow such a line, heating up, and soon reach the equilibrium line. Were it to contract further adiabatically, its pressure forces would overcome the gravitational forces and it would have to expand again. It would ultimately reside on the equilibrium line and contract no further.

Generally we find for an adiabat

$$T \propto \rho^{\gamma-1} \qquad (20.8)$$

For a molecular gas with 5 degrees of freedom we find for adiabatic contraction $T \propto \rho^{2/5}$. The temperature increases less steeply but the cloud will still soon reach the equilibrium line and stop contracting. For equation (20.4) to remain valid during contraction we must require that the temperature increases more slowly than

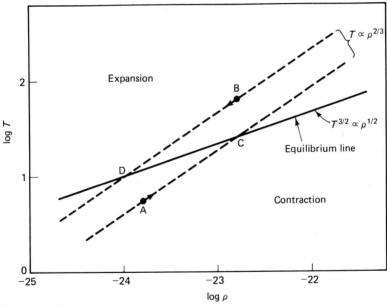

Fig. 20.1. In a temperature, density diagram we show schematically the line for which gravitational and pressure forces are in equilibrium, i.e., the line for which $T^{3/2} \propto \rho^{1/2}$ (solid line). Above this line the pressure forces are larger than the gravitational forces, leading to expansion. Below the line the inverse is true. Also shown are adiabats $T \propto \rho^{\gamma-1}$ for a monatomic gas with $\gamma = \frac{5}{3}$ (dashed lines). If a cloud at point A starts contracting adiabatically, it will reach the equilibrium line at point C and stop contracting. If a cloud at point B expands adiabatically it reaches the equilibrium line at point D and stops expanding.

$$T^{3/2} \propto \rho^{1/2} \tag{20.9}$$

For adiabatic contraction of monatomic gas as well as two atomic molecular gas the temperature increases too fast, the contraction would come to a halt immediately, unless the cloud can cool off by radiating enough to lose the excess energy. In Fig. 20.2 we have plotted again the equilibrium line for which $T \propto \rho^{1/3}$. We have also plotted adiabats for $\gamma = \frac{4}{3}$, for which also $T \propto \rho^{1/3}$. This line is parallel to the equilibrium line. For this γ a cloud at point A being unstable to gravitational contraction, and contracting adiabatically, would always remain below the equilibrium line and would therefore remain unstable to further gravitational contraction. This would also be the case if $\gamma < \frac{4}{3}$. Only for a polyatomic gas with more than 5 degrees of freedom could γ be less than $\frac{4}{3}$ and adiabatic contraction continue indefinitely. The formation of such molecules, however, requires heavy elements which are rare. Such polyatomic molecules, while they are observed in the interstellar medium, can generally form only in very small amounts. But we see that formation of molecules enhances the chances of gravitational contraction because of the increase in μ as seen in equation (20.7). γ could also become less than $\frac{4}{3}$ in clouds with temperatures around 10 000 K when hydrogen ionizes, as we saw

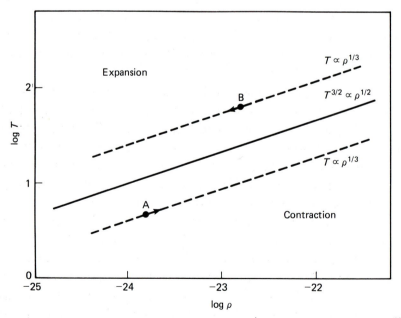

Fig. 20.2. Same as in Fig. 20.1, but adiabats for $\gamma = \frac{4}{3}$ are shown for which $T \propto \rho^{1/3}$. If a contracting cloud at point A follows this adiabat it will remain unstable to contraction. A cloud at point B will remain unstable to expansion.

when discussing the hydrogen convection zone. γ could also have values close to 1 for temperatures around 4000 K when hydrogen molecules dissociate, a process which requires a large amount of energy. A cloud with $\gamma < \frac{4}{3}$ always remains unstable against contraction if it starts from point A (Fig. 20.2) below the equilibrium line. However, it soon reaches temperatures for which ionization or dissociation is completed so that γ returns to larger values. It then follows the adiabat for the larger γ and reaches the equilibrium line again (see Fig. 20.1).

Of course, a cloud starting from point B (see Fig. 20.1) in the expansion region expands along an adiabat and would also reach the equilibrium line unless it has temperatures in the range where hydrogen molecules dissociate or where hydrogen ionizes.

In order to judge whether a gravitationally unstable interstellar cloud can continue to collapse we must study the possibilities of the cloud to radiate away the excess energy and to remain cool.

20.4   Non-adiabatic expansion and contraction of optically thin clouds

20.4.1 *Evolution of expanding clouds*

In reality the clouds do not expand or contract adiabatically; rather, they are in energy exchange with their surroundings. In order to see how their temperature actually changes we have to look at cooling mechanisms as well as heating processes. We follow here a qualitative discussion by Hayashi (1966).

In the interstellar gas, cooling is mainly due to atomic and molecular radiation and to dust radiation, which can become very important, especially for low gas temperatures and high densities. For high densities an interstellar cloud may become optically thick for all except infrared radiation, which is where the grains radiate. Then only this radiation can escape. For such clouds the grains are the most important coolants and can be expected to keep the cloud at a temperature around 10 K.

The interstellar gas and grains can also be heated by different mechanisms. Cosmic rays can penetrate the clouds and ionize the particles. Electrons with high energies are liberated in this process and may in turn ionize additional particles or transmit their kinetic energy to other particles by means of collisions. X-rays may also penetrate the clouds, ionizing atoms and liberating high energy electrons. If the clouds are not very dense the interstellar radiation field will penetrate the gas and may ionize particles. It will also excite some particles but this energy will be lost again in the cloud when a photon is reemitted and escapes into interstellar space.

We can describe the efficiency of cooling by the cooling times t_c, which are the times during which a cloud at given T and ρ would cool by a factor of e, that is

$$\frac{d \ln T}{dt} = \frac{1}{t_c} \qquad (20.10)$$

if we only consider the cooling processes. Here t stands again for time. Correspondingly we can describe the efficiency of heating by the heating time t_h. Heating is most efficient for small t_h. If heating is more efficient than cooling then

$$t_h < t_c \qquad (20.11)$$

In this case the cloud becomes warmer. It cools if

$$t_c < t_h \qquad (20.12)$$

In Fig. 20.3 we reproduce the results of calculations by Hayashi (1966). We show the domains in the T, ρ diagram where different heating or cooling mechanisms are dominating. Based on these estimates we can now also plot a line in the T, ρ plane where cooling is equal to heating. Clouds

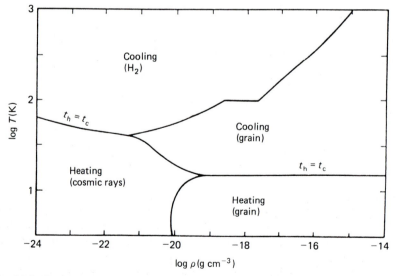

Fig. 20.3. In the ρ, T plane the domains are indicated in which different heating and cooling mechanisms dominate. Adapted from Hayashi (1966).

whose T and ρ combinations lie below this line heat up, those above this line cool. The temperatures and densities for which cooling equals heating are the equilibrium values for the interstellar medium.

In Fig. 20.4 we try to show what actually happens to gas clouds, taking into account these heating and cooling mechanisms, following Hayashi (1966). In order to judge the importance of cooling and heating we have to compare the heating and cooling times with the expansion or contraction times. If the heating times for instance are much shorter than the expansion times then any adiabatic cooling during expansion will be more than compensated immediately by the heating processes. Also if the cooling times are much shorter than the contraction times then the adiabatic heating during contraction will be more than compensated by the cooling processes. For a cloud which is considerably below the hydrostatic equilibrium line the gravitational forces are essentially unimpeded by the pressure forces and we can estimate the contraction time by equating it with the free fall time, t_f, which is given by

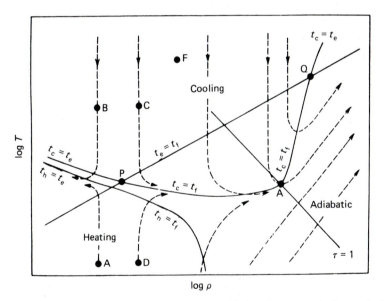

Fig. 20.4. In the ρ, T plane the hydrostatic equilibrium line $t_e = t_f$ is shown (solid line). Clouds with T above this line are unstable to expansion, for lower T they are unstable to contraction. Also shown are the domains in which the heating mechanisms dominate the cooling mechanisms, i.e., $t_h < t_c$, and those where cooling dominates heating, i.e. $t_c < t_h$. Also shown are the lines where the cooling times t_c equal the expansion times t_e and where they equal the free fall times t_f. Similarly the lines are shown where the heating times t_h equal the expansion times t_e and where they equal the free fall times t_f. Adapted from Hayashi (1966).

$$t_f = \frac{3\rho}{2} \bigg/ \frac{d\rho}{dt} = \left(\frac{32\pi G\rho}{3}\right)^{-1/2} \qquad (20.13)$$

If the pressure forces are much larger than the gravitational forces the cloud expands essentially with the velocity of sound. The expansion times, t_e, can then be estimated to be

$$t_e = \frac{R}{c_s} \qquad (20.14)$$

where c_s is the velocity of sound. On the equilibrium line the expansion time $t_e = t_f$, the free fall time. These times are important if we want to judge how fast heating or cooling work in comparison with contraction or expansion.

In Fig. 20.4 we have also plotted schematically the lines where the cooling time t_c equals the expansion time t_e and where it equals the free fall time t_f. Also shown are the lines on which the heating times t_h equal the expansion times and also where they equal the free fall times. With this information we can now follow what happens to a cloud which starts out way above the equilibrium line in the T, ρ plane of Fig. 20.4, say at point B. This point is above the equilibrium line and therefore in a domain where the cloud will expand. The cooling times are shorter than the heating times, so it will cool. Point B is also in a domain where the cooling times are shorter than the expansion times. This cloud therefore expands and cools not only because of the expansion but even more so because of the radiative energy losses. It cools very fast, essentially along a line of constant density because the expansion is relatively slow. When it crosses the line where $t_c = t_e$ the cooling by radiation becomes slower than the cooling due to the expansion, and it follows nearly an adiabat until it crosses the line where the heating times are equal to the expansion times. Below this line heating is more efficient than the adiabatic cooling. The cloud slowly increases its temperature while expanding. The cloud never reaches the equilibrium line, but keeps expanding.

We will now study what happens to a cloud which starts out at point C in the T, ρ diagram (Fig. 20.4). It has a higher density than the cloud at point B, but it is still in a domain in which it is unstable to expansion and where the cooling times are shorter than the expansion times. It also cools rapidly almost along a line of constant density. Because of its higher density it reaches the equilibrium while it is still in the domain of rapid cooling and it therefore continues to cool and crosses the equilibrium line. It now gets into a temperature, density range where the gravitational forces are larger than the pressure forces and starts to contract while still cooling. Were it to cross the line where the cooling time equals the free fall time ($t_c = t_f$) it would get into the domain of heating and would have to approach or even cross again the line where $t_c = t_f$. If it were to cross this line it would cool again. We see that this cloud will contract essentially along the line where $t_c = t_f$. Fig. 20.4 tells that in fact all the clouds which start out with densities larger than the one of cloud C will ultimately contract along the same line. They all will keep contracting even though they started out in a temperature, density domain where they were not unstable to gravitational contraction.

20.4.2 *Evolution of contracting clouds*

We shall now investigate what happens to clouds at positions A and D in Fig. 20.4. These clouds start out in the temperature, density domain where they are unstable to gravitational contraction. Let us first consider the cloud at position A. It is in the contraction domain and also in the heating domain. The heating time is much shorter than the free fall time. It heats up essentially at constant ρ, though slowly (relative to the heating) contracting until it crosses the equilibrium line, while still being in the heating domain. It will now stop contracting and instead start expanding while still increasing its temperature until it reaches the line where $t_h = t_c$. We can easily see that it cannot diverge much from this line but has to follow this line while expanding. This cloud started out in the gravitationally unstable regime, but due to the heating by cosmic rays and X-rays becomes too warm to continue its contraction.

We now follow a cloud starting out at position D in Fig. 20.4. It has a higher density than the cloud at position A. This cloud is also in the contracting and heating domain. It also follows a line of essentially, though not exactly, constant density. It first crosses the line where the heating time equals the free fall time. Beyond this line heating by cosmic rays or grain absorption becomes inefficient. The cloud contracts nearly along an adiabat until it reaches the line where $t_c = t_f$. If it were to cross this line it would have to cool and approach or cross this line again. This cloud also has to contract very close to the line where $t_c = t_f$. The same result is obtained for all clouds in the contraction domain with higher densities than the cloud at position D. All clouds which finally end up contracting will continue to do so along the line $t_c = t_f$. They will continue to contract as long as the cooling can keep up with the contractional heating.

20.5   Optically thick clouds and protostars

In the heating and cooling calculations discussed so far and on which Figs. 20.1 to 20.4 were based, it was assumed that the clouds are optically thin, which means that the cooling radiation can freely escape from the cloud. When the cloud contracts the optical depth along the radius does, however, increase and the cooling is reduced.

Why does the radial optical depth of a homogeneous contracting cloud increase? Suppose the absorption coefficient per gram of material κ_g remains constant during the contraction; then the radial optical depth τ through the cloud of radius R is given by (see Fig. 20.5)

$$\tau = \kappa_g \rho 2R \quad \text{and} \quad \rho = \frac{M}{\frac{4}{3}\pi R^3} \tag{20.15}$$

which means for a given mass

$$\tau = \kappa_g \frac{2 \times 3 \times M}{4\pi R^2} \propto \frac{M}{R^2} \tag{20.16}$$

Expressing this by means of the density rather than by means of the radius yields with $R = M^{1/3}/[(\frac{4}{3}\pi)^{1/3}\rho^{1/3}]$

$$\tau \propto \kappa_g \frac{2}{(\frac{4}{3}\pi)^{1/3}} M^{1/3}\rho^{2/3} \tag{20.17}$$

For each mass the radial optical depth of the cloud thus increases as $\tau \propto \rho^{2/3}$. For the decreasing radius there are more and more particles along the radial direction even for a given mass. (Of course for a given density the radial optical depths are larger for larger masses because the radius has to be larger.)

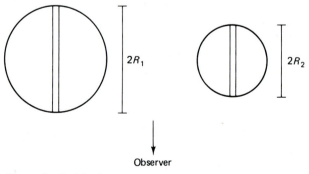

Observer

Fig. 20.5. For a cloud with given mass the number of particles along a column with cross-section 1 cm^2 and length $2R$ increases when the radius of the cloud decreases because $\rho \propto R^{-3}$ and therefore the optical depth of the cloud increases.

With the calculated κ_g for the interstellar medium we can for each cloud mass calculate the density for which the radial optical depth will become 1. At this point the cooling becomes inefficient, the cloud becomes hotter and κ_g increases rapidly with increasing T. The cloud quickly becomes optically thick. The line along which the contracting clouds are estimated to have an optical depth of 1 is indicated in Fig. 20.4. At this line the clouds start to contract nearly along an adiabat until they reach the hydrostatic equilibrium line. Contraction then ceases. The interstellar cloud has become a protostar in hydrostatic equilibrium contracting very slowly as discussed in Sections 2.3 and 10.4. We then have to consider the radiative transfer in the protostar. The timescale of contraction is determined by the time it takes the star to get rid of the excess gravitational energy gained during the contraction. For the star to remain in equilibrium half of the gravitational energy gain has to be radiated away before the star can continue to contract.

20.6   Fragmentation

In this section we shall discuss how low mass stars can possibly be formed. In order to better understand this we have in Fig. 20.6 redrawn part of Fig. 20.4 but have drawn boundary lines for gravitational instability for different masses, that is the equilibrium lines for different masses. We now follow the contraction line for a 1000 $M_\odot$ cloud, say, whose T and ρ correspond to point C in Fig. 20.4. Following the contraction line it will at point E cross the instability line for a 100 $M_\odot$ star. At this point it has a temperature and density that would make a 100 $M_\odot$ cloud unstable to contraction. If there are some minor inhomogeneities in the 1000 $M_\odot$ cloud, fractions of this cloud which have masses over 100 $M_\odot$ then may contract by themselves as 'subclouds' (see Fig. 20.7). We now follow the track of such a subcloud in Fig. 20.6. First, temperature and density are the same as in the original cloud. We can thus follow the same track as before in Fig. 20.6. When the subcloud contracts further it will at point F cross the instability line for a 10 $M_\odot$ cloud. This means that now even submasses of 10 $M_\odot$ become unstable to contraction. The subcloud may now itself break up into smaller subclouds. Upon further contraction these new subclouds will cross the instability lines for still smaller masses and then they may again break up into still smaller masses and form still lower mass clouds. Of course, this does not necessarily happen but it may happen. During the process of contraction the densities become high enough such that low mass subclouds become gravitationally unstable.

Of course, all the time the original mass of 1000 $M_\odot$ still keeps contracting – the whole cloud with all the subclouds could still collapse into one big supermassive 'star' if there is nothing to prevent it, such as, for instance, large-scale turbulence. Also, if the original big cloud has some small

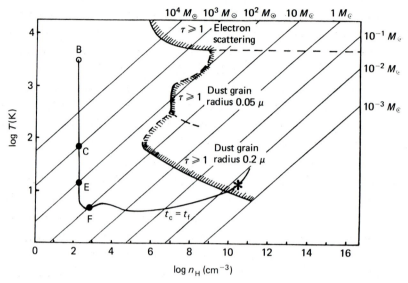

Fig. 20.6. In the T, $\log n_H$ plane (n_H = number of hydrogen atoms per cm³) hydrostatic equilibrium lines for different stellar masses are shown. Also shown is the contraction line for interstellar clouds starting at point B. When a cloud of $M = 1000\,M_\odot$ (point C) is unstable to contraction and contracts along the contraction line it crosses the hydrostatic equilibrium line for $M = 100\,M_\odot$ at point E. Beyond this line submasses of $M = 100\,M_\odot$ become unstable to contraction. If they contract further they will cross the equilibrium line for $M = 10\,M_\odot$ at point F. Beyond this point masses with $M = 10\,M_\odot$ become unstable to contraction, etc. Also shown is the relation between ρ and T for which clouds of different masses become optically thick (shaded curve). From Low and Lynden-Bell (1976).

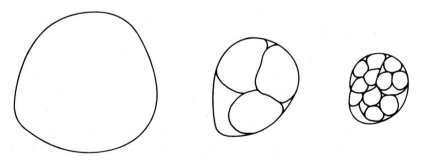

Fig. 20.7. During the contraction of a large mass with sufficient cooling the densities become high enough for subclouds to become gravitationally unstable. The large cloud may split into smaller subclouds. During further contraction still smaller subclouds become gravitationally unstable.

(or even very small) angular momentum around the central region this angular momentum is conserved during contraction. With decreasing radius the angular velocity ω must increase as $\omega \propto R^{-2}$ and the centrifugal forces F_c increase as $F_c \propto \omega^2 r \propto r^{-3}$; at some point these will prevent further contraction of the $1000\,M_\odot$ mass as a whole. This does, however, not prevent the subclouds from further contraction if there is turbulent motion in the cloud. The angular momentum around their contraction center may be much smaller. Eventually the centrifugal forces within the subcloud will also prevent the subclouds from further contraction and only still smaller parts with smaller angular momenta can contract further. In other words, the cluster as a whole cannot contract any further because of the centrifugal forces but the cloudlets leading to protostars may still be able to. Very massive stars are not expected to form if the subclouds with large masses have too much angular momentum. They have to reduce their angular momentum first, perhaps by turbulent viscosity, before they can contract further, or they may break up into smaller mass protostars before that happens. Rotation may perhaps enhance fragmentation.

There may be other mechanisms to prevent the massive cloud from further contraction. If for instance a massive star forms first in the center it may go through its evolution in 10^6 years, become a supernova and blow the cloud apart leading to large density fluctuations. The question is whether in such an event the high density regions can achieve a temperature low enough to lead to gravitational instability.

The fragmentation scenario would explain why young stars are preferentially observed in clusters and associations. Is there a limit to this fragmentation process or can planets be made like this? What is the smallest mass for a star to be formed by fragmentation? This question has been studied by several authors, first by Hoyle in 1953. If we could answer this question, we would know how far down the lower main sequence extends in the HR diagram. So far the smallest masses seen for stars are about $0.06\,M_\odot$ (for the stars Ross 614 B and Wolf 424 A). We observe that the number of stars increases steeply for smaller masses, down to masses of about $0.2\,M_\odot$. It therefore seems possible that there may be a large amount of mass in the low mass stars if there are many of them. So far hardly any have been found. Since they are so faint it is very difficult to observe them. It would therefore be nice if we could derive theoretically a lower mass limit for stars. Since the cooling is mainly due to dust and molecules it depends on the abundances of the heavy elements and we may perhaps expect that the lower mass limit on the main sequence depends somewhat on the element abundances. Studying the limiting magnitudes for the faint ends of the main sequences in globular clusters with different chemical abundances may provide clues or tests of our hypotheses. This is a very active field of research at the moment. It seems, however, that our instruments are not yet sensitive enough to answer this question (see Fig. 1.7).

20.7   Fragmentation limits

We see from Fig. 20.6 that fragmentation must stop when the fragments become optically thick and start contracting along the adiabats for which the temperature increase with increasing ρ is steeper than for the Jeans' mass, if $\gamma > \frac{4}{3}$ (see Fig. 20.1). For smaller masses the line $\tau = 1$ shifts to higher densities because for a given density the radius of the cloud becomes smaller for smaller mass. With $\tau = \kappa_g \rho R$ we find for $\tau = 1$ that

$$\rho R = \frac{1}{\kappa_g} = \rho^{2/3} M^{1/3} \left(\frac{3}{4\pi}\right)^{1/3} \tag{20.18}$$

For $\kappa_g = $ const. we find that for $\tau = 1$ we need $\rho \propto M^{-1/2}$. Knowing $\kappa_g(\rho, T)$ we can then along each mass line for contractional instability determine at which ρ and T (i.e., at which radius R) the radial optical depth through the cloud becomes $\tau = 1$. In Fig. 20.6 these points are connected by the shaded lines. For low T and larger ρ, or larger number of hydrogen atoms n_H per cm^3, the opacity is mainly due to dust grains with a radius around 0.2×10^{-4} cm. Following the contraction line $t_c = t_f$ it is obvious from this figure that there is a lower mass limit $M = M_F$ which can still become gravitationally unstable before becoming optically thick. This is the fragmentation limit. How large is M_F? According to Fig. 20.6, Low and Lynden-Bell (1976) find that $M_F \geq 0.007 \, M_\odot$. From an analytical discussion Rees (1976) finds a similar result. The uncertainty appears to be about a factor of 2. There seems to be no doubt that planets cannot be formed by fragmentation. They probably form by grains coalescing to form larger and larger particles, then rocks and finally planets.

Does the lower mass limit for fragmentation depend on the chemical composition? The instability boundary lines for the different masses remain unchanged. For smaller fractions of heavy elements the fraction of mass in grains must, however, be reduced. We therefore have less efficient cooling and smaller values of κ_g. The contraction line $t_c = t_f$ shifts to higher temperatures. The $\tau = 1$ line shifts to higher densities. From Fig. 20.6 it becomes clear that the two effects work against each other and that because of the coupling of κ_g and t_c no major changes in M_F may be expected.

Do we observe stars with masses close to M_F?

For decreasing masses the central temperatures are decreasing and the question is whether these low mass stars ever reach high enough temperatures in their interiors to start hydrogen burning. In other words: will they ever become main sequence stars or will they contract until they are degenerate and then cool down?

Present calculations show that stars with masses less than about 0.08 solar masses probably never start hydrogen burning. Stars with lower masses would be called brown dwarfs. The question is: do they exist? According to Low and Lynden-Bell's estimates the fragmentation limit is lower than this mass value, but so far no brown dwarfs have been detected. Perhaps they do not exist; in the fragmentation process such small masses may never become unstable for gravitational contraction because of some external influences, or if cooling is less efficient than assumed in Fig. 20.6.

20.8   Influence of magnetic fields

The discussions in the last sections have ignored the fact that we observe magnetic fields in the interstellar medium. These fields influence the contraction of a gravitationally unstable cloud. If the cloud is partially ionized (a very low degree of ionization is sufficient), then the gas becomes a very good conductor and the magnetic lines of force cannot slip through the gas (the magnetic field is 'frozen' in). The cloud can then only contract by also contracting the magnetic lines of force. This means the field strength is increased inversely proportional to the radius squared, i.e., $B \propto R^{-2}$. Will this prevent further contraction? In order to judge this we have to compare the magnetic pressure, which opposes the gravitational forces, with the gravitational forces, $F_g = GM_r/R^2$. These also increase proportional to R^{-2}. If in the beginning the gravitational forces are larger than the magnetic forces, they will remain larger during contraction. If, however, in the beginning the magnetic forces perpendicular to the magnetic lines of force are larger than the gravitational forces, no contraction can take place in this direction. The material can then only contract along the magnetic lines of force since in this direction there are no opposing forces from the magnetic field. The cloud then contracts to form a disk, which can fragment into smaller disks. With this different geometry all the studies discussed above will have to be revised. Cooling is more efficient for a thin disk than for a spherical cloud.

If the contracting cloud is so cool that there are essentially no free electrons and the gas is not a good conductor, then the magnetic field lines can slip through the material and contraction can take place unimpeded. For a 10 K cloud this is expected to be the case.

20.9   Position of protostars in the color magnitude diagram, Hayashi theory

We saw in Section 11.2 that no stars in hydrostatic equilibrium can appear to the right (i.e., on the cool side) of the Hayashi line in the color magnitude diagram. We therefore expect newly formed stars to first appear as stable stars somewhere on the Hayashi line. We also saw in the previous discussion that as long as $\gamma < \frac{4}{3}$ the star cannot reach an equilibrium position. As long as abundant elements like hydrogen or helium are in the process of ionizing over large fractions of the star we find $\gamma \leqslant \frac{4}{3}$ and the star cannot be stable. The star can therefore reach an equilibrium configuration only if it has become hot enough for most of its material to be ionized. The gravitational energy liberated in the collapse must have provided the ionization energy. Hayashi therefore estimated the radius of a star just reaching hydrostatic equilibrium by equating the liberated gravitational energy with the energy needed for ionization. This gives an estimate for the gravitational energy release at this point and thereby for the radius of a newly born star with a given mass. On the Hayashi line for this mass we can then find the star with the required radius. This determines the birth position of the star on its Hayashi line. For a star of one solar mass and homogeneous density and temperature Hayashi estimated that it would appear on the Hayashi line at a luminosity about 200 times the present solar luminosity but with a much lower temperature, of course, than the present sun. For smaller masses smaller luminosities are obtained. A group of newly born stars should fall along a given line in the HR diagram as shown in Fig. 20.8. We could call this the birth line of stars.

In Fig. 20.9 we show the positions of young T Tauri stars in the HR diagram. The coolest stars fall along a line not very different from that found by Hayashi, but we do not know which mass belongs to each observed point.

20.10   The initial mass function

What determines the initial mass function; that is, what determines the number of stars with a given mass to be born in a given molecular cloud? We do not know. We do know, however, that low mass stars contract slowly to the main sequence because they do not lose so much energy per unit of time. We saw in Chapter 2 that the contraction time goes as E_G/L where E_G is the gravitational energy release. With $E_G \propto GM^2/R$ and $L \propto M^{+3.5}$ we find for the gravitational contraction time (see Section 2.3) $t \propto 1/M^{1.5}R$.

Massive stars contract much more rapidly than lower mass stars, so they reach high surface temperatures rather quickly. Because of their strong radiation field they will probably blow off the surrounding dust cloud from which they formed by means of their radiation pressure and prevent more material from falling in. This may set an upper limit to the masses of stars which can be formed, even for very low rotation rates.

Once massive, hot stars are formed, their high radiation may well heat up the interstellar material in their surroundings, leading to temperatures too high to permit the formation of more low mass stars. It appears that low mass stars possibly form first. When the temperature in the cloud slowly increases because of these low mass stars, the fragmentation instability may stop at increasingly higher masses. Massive stars would then form last, but this is at present only speculation. We are far from understanding the mass distribution for newly formed stars.

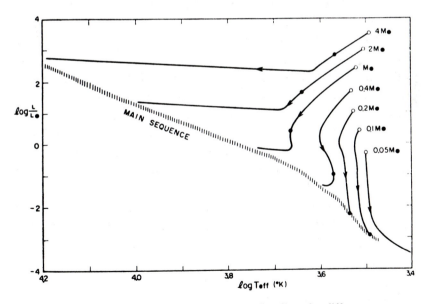

Fig. 20.8. In the log L, log T_{eff} diagram the contraction lines for different masses are shown, which first follow the Hayashi lines. The dots indicate the positions of homogeneous stars when they first arrive on the Hayashi line. From Hayashi (1966).

20.11   Inhomogeneous collapse of protostars

So far we have assumed that the collapsing cloud is a homogeneous entity with the same temperature and density everywhere. Actually, when a cloud contracts into a protostar we expect that close to the center higher densities are reached much faster than in the outer regions, because all the material is moving towards the center where it accumulates. Detailed calculations show that indeed the center reaches a quasi-equilibrium state much faster than the outer regions. When the next 'shells' of infalling material hit this quasi-equilibrium core shocks are generated which propagate back outwards, leading to higher temperatures in the outer layers for a short time. The paths of the different mass shells in the T, ρ diagram are rather complicated. They are even more complicated if rotation is included. We will not go into details here. A discussion may be found in the book *Protostars and Planets* (ed. Gehrels, 1978). Calculations show that while complicated phenomena occur in the center of the protostar we cannot see them because the interior is hidden from view by

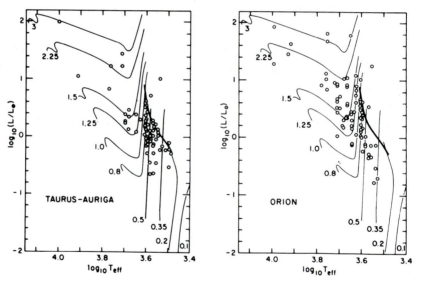

Fig. 20.9. The observed positions of young T Tauri stars in the Orion region and in the Taurus Auriga Association are shown in the $L, T_{\rm eff}$ diagram. Also shown are the Hayashi tracks for stars of different masses as given in the figure. A birth line for stars seems to be indicated. It is marked by a thick line. Hardly any stars are found on the cool side of this line. For a 1 $M_\odot$ star a birth luminosity of $\log L/L_\odot = 0.9 \pm 0.1$ seems to be indicated as compared to $\log L/L_\odot = 2.3$ estimated on Fig. 20.8. Of course, we do not know the masses of the observed T Tauri stars, which we would need to know for a quantitative comparison. The overall shape of the birth line resembles Hayashi's simple and rough estimate. From Stahler (1983).

the dust clouds in the outer layers of the protostar. These dust layers will ultimately be blown away, when the star has become hot enough to blow away the dust cocoon by radiation pressure on the grains. Calculations by Stahler (1983) indicate that the star should become visible on the Hayashi track with luminosities about a factor of 10 lower than estimated by Hayashi. This is mainly because the stars remain invisible longer, still embedded in the surrounding dust from which they accrete more mass.

20.12 Conclusion

While the comparison of observed HR diagrams and calculated evolutionary tracks tell us that we basically understand stellar structure and evolution, there are still many unsolved problems.

The prediction of the initial mass function for newly born stars is, of course, very important for the understanding of the history of globular clusters. This in turn is fundamental for the understanding of the enrichment of heavy element abundances in galaxies. For this, of course, we also have to know the details of stellar evolution, especially of massive stars, and the origin of the chemical elements. We have to know which stars become supernovae and how many heavy elements are generated in their interiors and which fraction of these elements is expelled into the interstellar medium.

All studies of the evolution of our own and other galaxies rely heavily on a good understanding of star formation and stellar evolution. Much has still to be learned.

Appendix

Radiative energy transport in stars

A.1 The transfer equation

We define the intensity I_λ as the amount of energy going per second and per wavelength interval $\Delta\lambda = 1$ cm perpendicularly through 1 cm^2 into the solid angle $\Delta\omega = 1$ (see Fig. A.1). We follow the changes of this intensity I_λ along its path s through the gas.

The intensity change due to absorption and emission is

$$\frac{dI_\lambda}{ds} = -\kappa_\lambda I_\lambda + \varepsilon_\lambda \tag{A.1}$$

Here ε_λ is the radiative energy emitted per unit volume per second per $\Delta\lambda = 1$ cm into the solid angle $\Delta\omega = 1$. (This ε_λ has nothing to do with the energy generation ε about which we talked in Chapter 8. ε_λ describes the emission of photons, following previous absorption processes.)

κ_λ is the absorption coefficient per cm at wavelength λ. It describes the fractional change of intensity after the beam has passed through 1 cm of gas.

Equation (A.1) is called the radiative transfer equation. The first term on the right-hand side describes the absorption, the second the emission from a column of 1 cm^2 cross-section per unit length $\Delta s = 1$ into a solid angle $\Delta\omega = 1$.

A black body is defined as a well-insulated box left without any energy exchange for a long time, such that everything is in complete thermodynamic equilibrium. For such a black body the emission ε_λ is given by

$$\varepsilon_\lambda = \kappa_\lambda B_\lambda \tag{A.2}$$

where the Planck function B_λ is

$$B_\lambda = \frac{2hc^2}{\lambda^5} \frac{1}{e^{hc/\lambda kT} - 1} \tag{A.3}$$

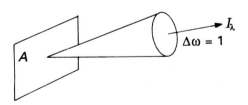

Fig. A.1. The intensity I_λ is defined as the amount of energy going per second perpendicularly through an area $A = 1$ cm^2 into a solid angle $\Delta\omega = 1$ in a wavelength band $\Delta\lambda = 1$.

and

$$B = \int_0^\infty B_\lambda d\lambda = \frac{\sigma}{\pi} T^4 \qquad (A.4)$$

Using relation (A.2) for the stellar interiors we find for the transfer equation

$$\frac{dI_\lambda}{ds} = -\kappa_\lambda I_\lambda + \kappa_\lambda B_\lambda \qquad (A.5)$$

The interior of a star almost resembles a black body. It is well insulated by all the mass around it and it has had a long time to establish equilibrium. We do not have complete thermodynamic equilibrium, though, because of the small energy flux F going through the star from the inside out; the insulation is not perfect. We can, however, easily estimate that this disturbance is minimal by comparing the energy flux F with the average intensity J of the radiation field.

A.2 The radiative flux

In order to determine the radiative flux F_r we have to calculate the net amount of energy going through an area A each second into a given direction. In our case only the radial flux is of interest. Because of the spherical symmetry there is no net energy flux flowing in the horizontal directions. The amount of energy dE_λ flowing through A per second into a direction making an angle θ with the normal $\vec{n}$ on A (see Fig. A.2) and going into the solid angle $d\omega$ is given by

$$dE_\lambda = A I_\lambda \cos \theta \, d\omega \qquad (A.6)$$

The total amount of energy going through A per second into the half sphere, πF_λ, is obtained by integrating over all solid angles $d\omega$. This means

$$A\pi F_\lambda = \int_\omega dE_\lambda = A \int_\omega I_\lambda(\varphi, \theta) \cos \theta \, d\omega \qquad (A.7)$$

Here we have indicated that in principle I_λ can depend on φ as well as on θ (see Fig. A.2). In the case of spherical symmetry there is, however, no reason why I_λ should depend on φ; it only depends on θ. In order to calculate the integral over $d\omega$ it is best to represent $d\omega$ by the area which this solid angle cuts out of the surface of a sphere with radius $R = 1$. For $d\omega$

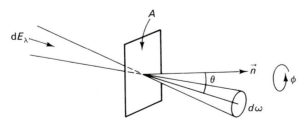

Fig. A.2. The amount of energy going through an area A under the angle θ with respect to the normal $\vec{n}$ on A into $\Delta\omega = 1$ is given by $A I_\lambda \cos \theta$.

corresponding to a change in φ by $d\varphi$ and to a change in θ by $d\theta$ (see Fig. A.3), we find that the surface area $d\sigma$ is given by $d\sigma = \sin\theta\, d\theta\, d\varphi = d\omega$. With this description of $d\omega$ we find that the whole solid angle covering all directions is given by the surface area of the unit sphere, namely $\omega = 4\pi$. We can easily verify this by integrating $d\omega$ over all angles $0 \leqslant \varphi \leqslant 2\pi$ and $0 \leqslant \theta \leqslant \pi$, which will cover the whole surface of the sphere, i.e.,

$$\omega = \int_{\text{whole surface}} d\omega = \int_{\varphi=0}^{2\pi} \int_{\theta=0}^{\pi} \sin\theta\, d\theta\, d\varphi$$

$$= 2\pi \int_{\theta=0}^{\pi} \sin\theta\, d\theta = 4\pi \int_0^1 \cos\theta\, d\theta = 4\pi \qquad (A.8)$$

With this description of $d\omega$ we can now also calculate πF_λ according to equation (A.7), namely

$$\pi F_\lambda = \int_\omega I_\lambda(\varphi, \theta)\cos\theta\, d\omega = \int_{\varphi=0}^{2\pi} \int_{\theta=0}^{\pi} I_\lambda(\varphi, \theta)\cos\theta\sin\theta\, d\theta\, d\varphi \qquad (A.9)$$

Since I_λ does not depend on φ in our spherically symmetric case the integral does not depend on φ and we can integrate over φ which gives a factor 2π.

$$\pi F_\lambda = 2\pi \int_{\theta=0}^{\pi} I_\lambda(\theta)\cos\theta\sin\theta\, d\theta \qquad (A.10)$$

In order to evaluate this integral we have to know $I_\lambda(\theta)$. In the case that $I_\lambda(\theta)$ does not depend on θ at all, i.e., if we have completely isotropic radiation with $I_\lambda(\theta) = \text{const.}$, we can take I_λ out of the integral and find

$$\pi F_\lambda = 2\pi I_\lambda \int_{\theta=0}^{\pi} \cos\theta\sin\theta\, d\theta = 2\pi I_\lambda \int_{\cos\theta=-1}^{\cos\theta=1} \cos\theta\, d(\cos\theta) = 0 \qquad (A.11)$$

A.3   Anisotropy of the radiation field in stellar interiors

The ratio of the flux

$$F_{\mathrm{r}} = \int_0^\infty F_\lambda\, d\lambda$$

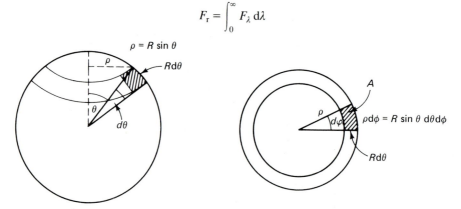

View from equatorial plane   View from direction of pole

Fig. A.3. The solid angle $d\omega$ is described by the area $d\sigma$ which the cone with opening $d\omega$ cuts out of the sphere with radius 1. For a cone with open $d\theta$ and $d\varphi$, $d\omega$ is given by $d\omega = \sin\theta\, d\theta\, d\varphi$.

to the average intensity, given by the Planck function B, is a measure for the anisotropy of the radiation field.

In layers with no energy generation we have $L(r) = \pi F_r 4\pi r^2 = $ const. At the surface where $r = R$ we find $\pi F_r = \sigma T_{\text{eff}}^4 = \pi F_r(R)$. In deeper layers it must therefore be

$$\pi F_r = \pi F_r(R) \frac{R^2}{r^2} = \sigma T_{\text{eff}}^4 \frac{R^2}{r^2} \qquad \text{(A.12)}$$

Assuming a temperature of 10^6 K at a radius $r = 0.8R$, and $T_{\text{eff}} = 8000$ K we find for instance

$$\frac{F_r}{B} = \frac{T_{\text{eff}}^4}{T^4} \frac{R^2}{r^2} \approx 6 \times 10^{-9}$$

The degree of anisotropy of the radiation field is extremely small as soon as the temperature T is much greater than T_{eff}. We can generally assume that the radiation field is isotropic, except in those cases for which the flux itself is important, as for instance for the energy transport.

This is very important when we now consider the radiative transfer in the spherically symmetric case.

A.4   The radiative transfer equation in stellar interiors

Let us go back to equation (A.1). In the plane parallel case of the stellar atmosphere (see Fig. A.4) we can replace the coordinate s along the path by the depth coordinate z. We find in this case

$$ds = -dz/\cos\theta \qquad \text{(A.13)}$$

and

$$\frac{dI_\lambda}{ds} = -\cos\theta \frac{dI_\lambda}{dz} = -\kappa_\lambda I_\lambda + \kappa_\lambda B_\lambda \qquad \text{(A.14)}$$

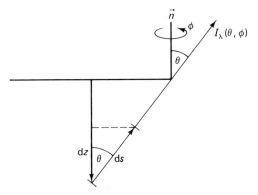

Fig. A.4. In the plane parallel atmosphere the coordinate s along the beam of light can be replaced by $s = -z/\cos\theta$.

where the minus sign indicates that dz points inwards, while ds points outwards. In the spherical case (see Fig. A.5) we find that the angle θ, which the light beam forms with the radius vector, changes along the path s; this means

$$s = s(r, \theta) \quad \text{and} \quad \frac{dI}{ds} = \frac{dI}{dr}\frac{dr}{ds} + \frac{dI}{d\theta}\frac{d\theta}{ds} \tag{A.15}$$

From Fig. A.6 we see that for small angles φ, i.e., small ds

$$x = r\, d\theta, \quad \cos\theta = \frac{dr}{ds} \quad \text{and} \quad \sin\theta = \frac{x}{ds} = \frac{r\, d\theta}{ds} \tag{A.16}$$

With these equalities we obtain from equations (A.13) and (A.1)

$$\frac{dI}{ds} = \frac{dI}{dr}\cos\theta + \frac{dI}{d\theta}\frac{\sin\theta}{r} = -\kappa_\lambda I_\lambda + \kappa_\lambda B_\lambda \tag{A.17}$$

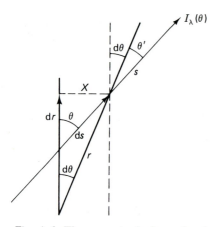

Fig. A.5. In the spherical case the angle θ which the beam of radiation forms with the normal $\vec{n} = \vec{r}$ on the 'horizontal' layer changes along the beam of light.

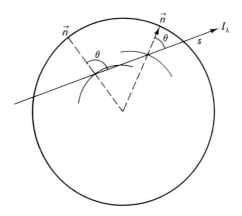

Fig. A.6. The geometry is shown for the spherical case. Along the path s of the light beam the angle θ between the light beam and the radius vector changes from θ to θ'. For small values of ds and dθ the relations (A.16) hold.

For isotropic radiation $dI/d\theta = 0$ and we obtain

$$\cos\theta \frac{dI_\lambda}{dr} = -\kappa_\lambda I_\lambda + \kappa_\lambda B_\lambda \qquad (A.18)$$

which is the same as equation (A.14) when we replace dz by $-dr$. Of course, this is only true for an isotropic radiation field. In this case the spherical transfer equation is the same as that for the plane parallel case. For the stellar interior this is a very good approximation, as we saw.

In Chapter 2 we calculated that in stellar interiors the mean free path for photons is about 1 cm. This is much smaller than the radius R. This is, of course, the reason why spherical symmetry is unimportant for radiative transfer in stellar interiors. Because of this the radiation field is very nearly isotropic.

A.5   Relation between temperature stratification and radiative flux

We now multiply equation (A.18) by $\cos\theta$ and integrate over the whole sphere, which means we integrate over the solid angle ω. In order to integrate at a given r over the whole solid angle we have to integrate over φ from 0 to 2π and over θ from 0 to π. Integration of equation (A.18) over the whole solid angle after multiplication with $\cos\theta$ leads to

$$\int_0^{2\pi} \int_0^\pi \frac{dI_\lambda}{dr} \cos^2\theta \sin\theta \, d\theta \, d\varphi$$

$$= -\kappa_\lambda \int_0^{2\pi} \int_0^\pi I_\lambda(r,\theta,\varphi)\cos\theta\sin\theta\,d\theta\,d\varphi + \kappa_\lambda \int_0^{2\pi} \int_0^\pi B_\lambda(r)\cos\theta\sin\theta\,d\theta\,d\varphi$$

$$= -\kappa_\lambda \pi F_\lambda(r) + 0 \qquad (A.19)$$

Here we have made use of the fact that $B_\lambda(r)$ is isotropic and can therefore be taken out of the second integral on the right-hand side. We are then left with

$$B_\lambda(r) \int_0^{2\pi} \int_0^\pi \cos\theta\sin\theta\,d\theta\,d\varphi = 0 \qquad (A.20)$$

For the left-hand integral we have again made use of the fact that $I_\lambda(r,\theta,\varphi)$ is essentially isotropic and therefore dI_λ/dr is independent of the angle θ. We find then from equation (A.19) that

$$\frac{dI_\lambda}{dr} \int_0^{2\pi} \int_0^\pi \cos^2\theta\sin\theta\,d\theta\,d\varphi = \frac{2\times 2\pi}{3}\frac{dI_\lambda}{dr} = -\kappa_\lambda \pi F_\lambda(r) \qquad (A.21)$$

Since the stellar interior is almost a black body, we have $I_\lambda \approx B_\lambda$. We now divide equation (A.21) by κ_λ and integrate over all wavelengths λ. This yields

$$\frac{4}{3} \int_0^\infty \frac{1}{\kappa_\lambda} \frac{dB_\lambda}{dr} d\lambda = -\int_0^\infty F_\lambda(r)\,d\lambda = -F_r(r) \qquad (A.22)$$

where $F_r(r)$ is the radiative energy flux.

If we further assume κ_λ to be independent of the wavelength λ, i.e., if we make the so-called grey approximation, we can write

$$\frac{4}{3} \int_0^\infty \frac{1}{\kappa_\lambda} \frac{dB_\lambda}{dr} d\lambda = \frac{4}{3}\frac{1}{\kappa}\frac{dB}{dr} = -F_r(r) \qquad (A.23)$$

If κ_λ is not independent of λ, as we assumed deriving equation (A.23), we can still derive the same relation with a properly determined mean value of κ_λ, which turns out to be the Rosseland mean absorption coefficient κ_R (see also Volume 2).

A.6   The Rosseland mean absorption coefficient

In deriving equation (A.23) we should have calculated

$$\int_0^\infty \frac{1}{\kappa_\lambda} \frac{dB_\lambda}{dr} d\lambda$$

which we want to replace by

$$\int_0^\infty \frac{1}{\kappa_\lambda} \frac{dB_\lambda}{dr} d\lambda = \frac{1}{\bar\kappa} \frac{dB}{dr}$$

For equation (A.23) to hold we must therefore require that

$$\int_0^\infty \frac{1}{\kappa_\lambda} \frac{dB_\lambda}{dr} = \frac{1}{\bar\kappa} \int_0^\infty \frac{dB_\lambda}{dr} d\lambda \tag{A.24}$$

With

$$\frac{dB_\lambda}{dr} = \frac{dB_\lambda}{dT} \frac{dT}{dr} \tag{A.25}$$

and with dT/dr being independent of λ we derive

$$\frac{1}{\bar\kappa} \frac{dB}{dT} = \int_0^\infty \frac{1}{\kappa_\lambda} \frac{dB_\lambda}{dT} d\lambda \quad \text{or} \quad \frac{1}{\bar\kappa} = \frac{\int_0^\infty \dfrac{1}{\kappa_\lambda} \dfrac{dB}{dT} d\lambda}{\dfrac{dB}{dT}} \tag{A.26}$$

as the prescription to determine the proper mean value of κ_λ. This $\bar\kappa$ is a harmonic average and is called the Rosseland mean absorption coefficient κ_R. Equation (A.24) tells us that the appropriate mean of κ_λ must weigh most heavily those parts of the spectrum where the flux

$$F_\lambda \propto \frac{1}{\kappa_\lambda} \frac{dB_\lambda}{dr}$$

is large, which are those wavelengths for which κ_λ is small and B_λ is not small.

Problems

Chapter 1

1. For a star whose spectrum indicates that it is an A0 V star (strong H lines) the measured color is $B - V = 0.20$, and the apparent magnitude $m_V = 9.6$. How large is m_B? How large is $E(B - V)$? What is $(B - V)_0$, $(U - B)_0$, m_{V_0} and m_{B_0}? How large is $U - B$? What is the distance modulus $m_{V_0} - M_V$? How far away is the star?

2. After correction for the interstellar extinction stars with $(B - V)_0 = 0.6$ in the Praesepe star cluster have an apparent visual magnitude of $m_{V_0} = 10.7$. Determine the distance to the star cluster.

Chapter 2

1. Determine the isothermal scale height for the earth's atmosphere assuming $T = 300$ K. $M(\text{earth}) = 5.98 \times 10^{27}$ g, $2\pi R(\text{earth}) = 40\,000$ km. $G = 6.68 \times 10^{-8}$ [cgs]. Compare this with the isothermal scale height for the solar atmosphere. Use $T_\odot = 5800$ K, $M_\odot = 2 \times 10^{33}$ g, $R_\odot = 6.98 \times 10^{10}$ cm. Do the same for the solar interior. For a rough estimate you can use the atmospheric $g_\odot$ and $T \sim 10^7$ K for an interior layer.

2. At the bottom of an isothermal atmosphere the gas pressure is $P_g = P_{g_0}$ and the density $\rho = \rho_0$. Calculate how much mass is above this layer.
   You now replace in your mind the atmosphere with a layer of gas with constant density ρ_0. What is the height of this hypothetical atmosphere if it has the same mass as the real atmosphere?

3. Estimate the diffusion time for photons to get to the surface from a layer in the sun which is $0.1\,R_\odot$ below the surface, and has a density of $\rho = 2 \times 10^{-2}$ g cm^{-3} and a κ per particle which is 2×10^{-22} cm^2.

4. Estimate the central pressure and temperature in an O star with $M = 40\,M_\odot$ and $R = 30\,R_\odot$.

5. In a supergiant of solar temperature the gas pressure P_g in the atmosphere is $P_g(\tau = \frac{2}{3}) \sim 3 \times 10^3$ dyn cm^{-2}. How thick is the atmosphere? Assume $\kappa_g = 1$ (not quite true). How large is the mean free path of the photons?

6. For most stars (all main sequence stars) the ratio of mass M to radius R changes very little (at most about 50 per cent for $T_{\text{eff}} > 4000$ K). This means $M/R \sim$ constant. Assume that for all stars the average cross-section for photon absorption is the same. Calculate $\bar{\rho}$ for a star with $R = 10\,R_\odot$ and calculate the time which the photon needs to get from the interior to the surface for such a star. Compare this with the time needed for the photons in the sun. $M_\odot = 2 \times 10^{33}$ g, $R_\odot \approx 7 \times 10^{10}$ cm.

7. For which equatorial rotational velocities do the centrifugal forces become 10 per cent of the gravitational forces? Consider main sequence stars for which the gravitational acceleration $g \approx 10^4$ cm s^{-2}. Make calculations for $R = R_\odot = 700\,000$ km and $R = 10\,R_\odot$.

8. For a homogeneous magnetic field the magnetic forces can be described by the magnetic pressure $P_m = H^2/8\pi$. (H = magnetic field strength in Gauss.) Estimate for which magnetic field strengths magnetic forces may become more important than gas pressure forces. In the sun $P_g \sim 10^5$ dyn cm^{-2}, in hot stars $P_g \sim 10^4$ dyn cm^{-2} in the atmospheres. Could magnetic forces become more important in stellar interiors?

9. Calculate the energy needed to completely ionize the stellar material consisting of 91 per cent hydrogen and 9 per cent helium. Compare with the kinetic energy for $T \sim 10^7$ K. $\chi_{ion}(H) = 13.6$ eV, $\chi_{ion}(He) = 24.5$ eV, $\chi_{ion}(He^+) = 54.4$ eV. 1 eV = 1.66 $\times$ 10^{-12} erg.

Chapter 3

1. Most stars apparently do not change their radii. We concluded they must be in hydrostatic equilibrium. Suppose hydrostatic equilibrium were violated by 0.01 per cent, which means a fraction of 10^{-4} of the gravity would be imbalanced by the pressure gradient. How long would it take for the sun to change its radius by 10 per cent? $R_\odot = 7 \times 10^{10}$ cm, $g_\odot = 2.74 \times 10^4$ cm s^{-2}.

2. If thermal equilibrium were violated by 0.01 per cent, which means the heat transport coming into the atmosphere from the bottom would be $(1 - 10^{-4})$ times the flux going out at the surface, how long would it take for the sun to change its temperature by 10 per cent? The gas pressure in the solar atmosphere is $P_g \approx 10^5$ dyn cm^{-2}. $T_{eff} = 5800$ K.

3. Estimate the time span during which the O star atmosphere would cool off if no heat were supplied from below. Assume a temperature of 45 000 K for the O star and $P_g = 10^3$ dyn cm^{-2}. The atmosphere has an extent of about 1 pressure scale height H. Assume $g = 10^4$ cm s^{-2}.

4. In the solar atmosphere κ per particle is about 10^{-24} cm^2. The gas pressure P_g is roughly $P_g \sim 10^5$ dyn cm^{-2}, the temperature $T \sim T_{eff} = 5800$ K. The gas is mainly hydrogen. Calculate the radiative flux πF_r. Estimate by how much τ changes over one pressure scale height H.

5. Compare the conductive heat flux for the solar atmosphere with the radiative energy flux.

Chapter 4

1. Calculate the Rosseland mean opacity for a so-called picket fence model of the absorption coefficient (see Fig. B.1). Assume that over 10 per cent of the wavelengths κ_g is 100 cm^{-2}, while over 90 per cent of the wavelengths $\kappa_g = 1$ cm^2, for all wavelength regions.

2. Calculate λ_{max} for $T = 3500$ K, 6000 K, 10 000 K, 40 000 K.

3. Calculate the bound-free continuous absorption coefficient for hydrogen including only the Balmer level with $n = 2$ and the Lyman level with $n = 1$. Plot $\log \kappa_{at} = \kappa$ per atom as a function of $\log \lambda$ or $\log \nu$. The absorption coefficient per electron in the quantum state n is given by

$$a_n = \frac{64\pi^4 m_e e^{10} Z'^4 g_G}{3\sqrt{3}\, ch^6 n^5 \nu^3}$$

Take into account

$$\frac{N(n = 2)}{N(n = 1)} = \frac{g(n = 2)}{g(n = 1)} \exp\left(-\chi_{\mathrm{exc}}/kT\right)$$

Use $g_G = 1.$ $m_e = 9.105 \times 10^{-28}$ g, $c = 3 \times 10^{10}$ cm, $h = 6.624 \times 10^{-27}$ ergs, $e = 4.8024 \times 10^{-10}$ electrostatic units, $g_G =$ Gaunt factor.

Fig. B.2 shows the schematic energy level diagram of the hydrogen atom including only the two lowest energy levels.

Chapter 5

1. Calculate the radiative gradient ∇_r for depth independent κ_g. By which factor must C_V increase in order to render such an atmosphere convectively unstable?

2. For two-atomic molecules there are two additional degrees of freedom because of rotation. How large is C_V? How large is ∇_{ad} in a stellar gas layer with two atomic molecules? For a depth independent κ_g would the gas layer be unstable to convection?

3. Assume $\gamma = \frac{5}{3}$ and $\kappa_g = \kappa_0 P_g^b$. How large must be the value of b in order to find convective instability?

4. If half of the energy is transported by radiation, how large is the temperature gradient ∇ as compared to ∇_{rad}?

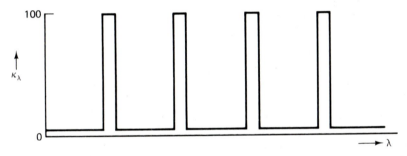

Fig. B.1. The wavelength dependence of κ_λ for a picket fence model is shown. This distribution of small and large κ_λ extends over the whole spectral region.

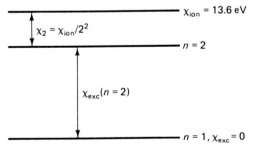

Fig. B.2. Shows the schematic energy level diagram of the hydrogen atom including only the two lowest energy levels and the ionization limit.

Chapter 6

1. Calculate the convective energy flux in the solar photosphere where $\Delta T \sim 300$ K, $P_g \sim 10^5$ dyn cm^{-2}, $T \approx 5800$ K, $v \sim 2$ km s^{-1}. $C_V(\text{mol}) = \frac{3}{2}R_g$. Compare this with the radiative flux.

Do the same for an early F star where $P_g \sim 10^3$ dyn cm^{-2}, $T_{\text{eff}} \sim 7500$ K, $\Delta T \sim 500$ K, $v \sim 3$ km s^{-1}. $\sigma = 5.6 \times 10^{-5}$ [cgs].

2. Estimate the maximum convective velocity in an atmosphere of a K star.

3. Calculate the ratio of radiative to convective flux in a layer with adiabatic temperature stratification at the top of the solar convection zone where $\nabla_r \sim 1000$ and $\nabla_{\text{ad}} \sim 0.2$, and close to the bottom of the solar convection zone where $\nabla_r \sim 0.5$ and $\nabla_{\text{ad}} \sim 0.35$.

Chapter 7

1. Calculate and plot $P_g(T)$ for which in a pure hydrogen atmosphere the ratio $(n(\text{H}^+))/(n(\text{H})) = 10$. Use the Saha equation

$$\frac{n(\text{H}^+)}{n(\text{H})} = \frac{1}{n_e}\frac{u^+}{u}2\frac{(2\pi m_e kT)^{3/2}}{h^3}e^{-13.6eV/kT}$$

Remember: $n(\text{H}^+) = n_e$ in a pure hydrogen atmosphere, $P_g = (n(\text{H}) + n(\text{H}^+) + n_e)kT$, $u^+ = 1$, $u = 2$.

Suggested procedure: Choose $n_e \geqslant 10^{17}$, calculate T for which $(n(\text{H}^+))/(n(\text{H})) = 10$, then calculate P_g for this T and n_e. Calculate five points with $\Delta \log n_e \sim 1$.

Chapter 8

1. Calculate how long an A0 star can live on its gravitational energy. An A0 main sequence star has $M/M_\odot \sim 2.5$ and $R/R_\odot \sim 2.0$. Its T_{eff} is about 10 000 K.

2. Suppose the sun were made of the most favorable proportions of H and O such that all the mass could burn to H_2O. How long could the sun live on such energy supply? Assume that for each H_2O molecule an energy of about 10 eV is liberated.

3. Calculate how long an A0 main sequence star can live on the main sequence on its nuclear energy source. Use the data given in problem 1.

4. Estimate the age of the globular cluster 47 Tuc from the colors and the luminosities of the stars just evolving off the main sequence (see Fig. 1.7). Assume that the 47 Tuc main sequence fits exactly on the main sequence of the stars in our neighborhood. (This is actually not true because 47 Tuc is metal poor. Its main sequence lies below the one for stars with solar element abundances. In addition the stars are more blue than solar neighborhood stars because less light is absorbed in the blue in the spectral lines of the heavy elements. See Volume 2.)

Give another age estimate assuming that in the color absolute magnitude diagram the 47 Tuc main sequence lies 0.8 magnitude below the main sequence of the solar neighborhood stars (see Fig. 14.14).

Chapter 9

1. Verify equations (9.15) to (9.19).

Chapter 10

1. Assume that all main sequence stars have a helium abundance by mass of $Y = 0.28$. How much helium enrichment does it take to increase L by a factor 1.5 and thereby reduce the main sequence lifetime of the stars by a factor of 1.5? (a) for B stars, (b) for G stars.

2. Derive the μ dependence of the luminosity if we had $\kappa_g = \kappa_0 \rho^{0.5} T^{-2}$. Would a mixed star evolve above or below the main sequence?

3. In F stars the energy generation is partly due to the proton–proton chain and partly due to the CNO cycle. Assume that on average $\nu \sim 10$. Which way would a well mixed star evolve?

Chapter 11

1. Show the change in the position of the Hayashi line for a change in mass by a factor of 2 as compared to the Hayashi line for $M = 1\ M_\odot$. Calculate $\Delta \log T_{\text{eff}}$ for a given luminosity $L \leq L_\odot$.

2. Show the change in position of the Hayashi lines for $\log Z/Z_\odot = -1$ and $\log Z/Z_\odot = -2$ as compared to $\log Z/Z_\odot = 0$. Calculate $\Delta \log T_{\text{eff}}$ for a given $L \leq L_\odot$.

Chapter 12

1. Calculate the values of M_r/M for equal spacing of

$$\xi = \ln \left(1 - \frac{M_r}{(1 + \eta)M} \right)$$

in steps of 0.05, $\xi = 0$ to 1. Use $\eta = 0.01$ and $\eta = 0.001$.

2. Calculate the M_r values for $\xi_{j+1/2}$ and compare with $(M_{rj} + M_{rj+1})/2$ for $\xi = 0.1$ and $\xi = 0.5$. How important is it to center correctly at $\xi_{j+1/2}$?

3. Suppose you had chosen $\ln M_r$ as your independent variable. You then have to determine the pressure etc. at $(\ln M_r)_{j+1/2}$. How large an error do you make if you use $(M_r)_{j+1/2}$ instead? Estimate your errors by looking at Table 13.2 which gives the relation between M_r, P_g and $T(r)$ for a star with $M = M_\odot$.

4. Verify equations (12.9), (12.10) and (12.12).

Chapter 13

1. Determine M_{bol} and $B - V$ for the zero age sun and for the present sun from the data given in Tables 13.1, 13.2 and 1.1.
   Do the same for the zero age star with 15 $M_\odot$ (Table 13.3) and the same star after it has been on the main sequence for 8.6×10^6 years. Plot the data in the color magnitude diagram.

2. Calculate the equilibrium abundance ratio for C^{13}/C^{12} and N^{14}/C^{12} in the center of the present sun. Use the reaction times quoted in equations (8.9).

3. Plot temperature and pressure for the B0 star as a function of M_r for the zero age main sequence star and for the evolved main sequence star. Discuss what happens to the different mass shells and why. Use Tables 13.4 and 13.5.

4. Plot L/L^* as a function of M_r/M^* for the star with $T_{eff} = 20\,423$ K star and for the sun. Discuss and explain the differences. Use Tables 13.1 and 13.4.

5. Plot $T(r)$, $P_g(r)$ and the chemical abundances as a function of M_r/M for the stars with $T_{eff} = 10\,800$ K stars. Use Tables 13.5 and 13.6. Discuss the evolutionary changes.

6. Calculate T_{eff} from L and R for the models in Tables 13.3 to 13.6.

Chapter 14

1. Estimate the neutron densities for which neutron degeneracy becomes important and also where relativistic neutron degeneracy occurs.

Chapter 15

1. Calculate the lifetime of a pure helium star with $M = 3\,M_\odot$ on the helium main sequence if it burns 10 per cent of its helium to carbon by means of the triple-alpha reaction. Use the luminosities given in Fig. 10.7.

Chapter 16

1. Calculate the cooling sequence for the star Sirius B which now has $m_v = 8.68$ and $T_{eff} = 25\,500$ K. Sirius has a trigonometric parallax of $\pi = 0\rlap{.}''377$. Use the bolometric correction given in Table 1.1 for main sequence stars. Calculate its apparent magnitudes m_v as a function of its T_{eff}.

2. Estimate roughly the apparent magnitude of a neutron star at the distance of Sirius if it also had $T_{eff} = 25\,500$ K, and if it had $T_{eff} = 10^5$ K.

Chapter 17

1. Assume that the pulsation constants Q are the same for all stars (not true but a reasonable approximation for order of magnitude estimates). If main sequence stars were radial pulsators, what would be the fundamental periods for the sun, for Vega, for the B0 V star τ Sco? What would be the period for a white dwarf of 0.6 $M_\odot$ with a radius of roughly 6000 km? Assume $Q = 0.04$ days.

Chapter 18

1. Plot a luminosity, T_{eff}, diagram for the Praesepe cluster using the observed color magnitude diagrams (Fig. 1.5) and Table 1.1. Assume $E(B - V) = 0.0$ and $m_{V_0} - M_V = 6.0$.

Chapter 19

1. Calculate the Jeans mass for a gas with $n = 10^{-1}$ cm^{-3} and $T = 100$ K, and with $T = 1000$ K. Do the same for $n = 10$ cm^{-3}.

2. Verify equation (20.13).

3. Follow the evolution of a cloud at point F in Fig. 20.4. Will it end up contracting or expanding?

4. For a mass of 10 $M_\odot$, at which radius does a spherically symmetric homogeneous hydrogen cloud reach a density of $\log n_H = 6$ and become optically thick if it has a temperature of 50 K?

5. (a) Suppose an interstellar cloud is penetrated by a homogeneous magnetic field H. If the cloud has $n = 10^4$ cm^{-3} and $T = 10$ K, for which field strength does the gas pressure P_g become equal to the magnetic pressure $P_H = H^2/8\pi$ (H in Gauss)?

(b) If a cloud of the Jeans mass for this density and temperature and with half this field strength H contracts to the main sequence with a frozen-in magnetic field, how large would the field strength H^* be for the main sequence star if it had homogeneous density?

References

Arp, H. C. 1958, in *Handbook of Physics*, vol. 51, Springer Verlag, Berlin, p. 75.

Bahcall, J. and Ulrich, R. 1987. Preprint.

Baker, N. and Kippenhahn, R. 1962, *Zeitschrift für Astrophysik*, **54**, 114.

Becker, S. A., Iben, I. and Tuggle, R. S. 1977, *Astrophys. J.*, **218**, 633.

Becker, W. 1950, *Stern und Sternsystem*, 2nd edn, Theodor Steinkopff, Dresden.

Bertelli, G., Bressan, A. G. and Chiosi, C. 1984, *Astron. Astrophys.*, **130**, 279.

Biermann, L. 1937, *Astron. Nachrichten*, **264**, 359.

Boesgaard, A. M. and Tripicco, M. J. 1986, *Astrophys. J.*, **308**, L49.

Böhm-Vitense, E. 1953, *Zeitschrift f. Astrophysik*, **32**, 135.

Böhm-Vitense, E. 1958, *Zeitschrift f. Astrophysik*, **46**, 108.

Brunish, W. M. 1989, private communication.

Brunish, W. M. and Truran, J. W. 1982, *Astrophys. J. Suppl.*, **49**, 447.

Burnham, R. 1978a, *Burnham's Celestial Handbook*, vol. III, Dover Publ., New York, p. 1442.

Burnham, R. 1978b, *Burnham's Celestial Handbook*, vol. III, Dover Publ., New York, p. 1446.

Burnham, R. 1978c, *Burnham's Celestial Handbook*, vol. III, Dover Publ., New York, p. 1911.

Cohen, M. and Kuhi, L. V. 1979, *Astrophys. J. Suppl.*, **41**, 743.

Cox, J. P. 1956, *Astrophys. J.*, **122**, 286.

Cox, J. and Salpeter, E. E. 1964, *Astrophys. J.*, **140**, 485.

Cox, A. N. and Tabor, J. E. 1976, *Astrophys. J. Suppl.*, **31**, 271.

Eggleton, P. P. 1971, *Monthly Notices Roy. Astr. Soc.*, **151**, 351.

Faulkner, J., Griffiths, K. and Hoyle, F. 1965, *Monthly Notices*, **129**, 363.

Finkelnburg, W. and Peters, Th. 1957, in *Handbook of Physics*, vol. 28, Springer Verlag, Berlin, p. 79.

Gehrels, T. 1978, *Protostars and Planets*, University of Arizona Press, Tucson, Arizona.

Hayashi, C. 1966, *Ann. Rev. Astron. Astrophys.*, **4**, 171.

Hayashi, E., Hoshi, R. and Sugimoto, D. 1962, *Supplement of the Progress of Theoretical Physics*, no. 22.

Henyey, L. G., Wilets, L., Böhm, K. H., Le Levier, R. and Levee, R. D. 1959, *Astrophys. J.*, **129**, 628.

Hesser, J. E., Harris, W. E., VandenBerg, D. A., Allwright, J. W. B., Shott, P. and Stetson, P. B. 1987, *Publ. Astr. Soc. Pacific*, **99**, 739.

Hoyle, F. 1953, *Astrophys. J.*, **118**, 512.

Iben, I. 1967, *Ann. Rev. Astron. Astrophys.*, **12**, 215.

Iben, I. 1971, *Publ. Astr. Soc. Pacific*, **83**, 697.

Iben, I. and Tuggle, S. 1972, *Astrophys. J.*, **178**, 441.

Johnson, H. L. and Morgan, W. W. 1953, *Astrophys. J.*, **117**, 313.

Kippenhahn, R., Weigert, A. and Hofmeister, E. 1967, in *Methods of Computational Physics*, vol. 7, p. 129.

Kurucz, R. 1979, *Astrophys. J. Suppl.*, **40**, 1.

Lambert, D. L. and Ries, L. M. 1981. *Astrophys. J.*, **248**, 228.

Ledoux, P. and Walraven, Th. 1958, in *Handbook of Physics*, vol. 51, Springer Verlag, Berlin, Göttingen, Heidelberg, p. 353.

Low, C., Lynden-Bell, D. 1976, *Monthly Notices Roy. Astr. Soc.*, **176**, 367.

Mateo, M. 1987, Ph.D. Thesis, University of Washington.

O'Dell, R. 1968, in *Planetary Nebulae* (IAU Symp. No. 34), D. E. Osterbrock and C. R. O'Dell (eds.), p. 361, Dordrecht: Reidel.

Popper, D. M. 1980, *Ann. Rev. Astron. Astrophys.*, **18**, 115.

Proffitt, C. 1988, Ph.D. Thesis, University of Washington.

Rees, M. 1976, *Monthly Notices Roy. Astr. Soc.*, **176**, 483.

Rosseland, S. 1949, *The Pulsation Theory of Variable Stars*, Clarendon Press, Oxford.

Sandage, A. R. 1986, *Ann. Rev. Astron. Astrophys.*, **24**, 450.

Sandage, A. R. and Tammann, G. 1968, *Astrophys. J.*, **151**, 531.

Schmidt, E. 1984, *Astrophys. J.*, **285**, 501.

Stahler, S. W. 1983, *Astrophys. J.*, **274**, 822.

Stetson, P. B., VandenBerg, D. A., Bolte, M., Hesser, J. E. and Smith, G. H. 1989, *Astron. J.*, **97**, 1360.

Schwarzschild, M. 1958a, *Stellar Structure and Evolution*, p. 72.

Schwarzschild, M. 1958b, *Stellar Structure and Evolution*, p. 82.

Schwarzschild, M. 1958c, *Stellar Structure and Evolution*, p. 54.

Spite, F. and Spite, M. 1982, *Astr. Astrophys.*, **115**, 357.

Straizys, V. and Sviderskiene, Z. 1972, *Vilniaus Astronomijos Observatorijos Biuletenis*, **35**, 1.

Ulrich, R. K. 1974, *Zeitschrift für Astrophys. J.*, **188**, 369.

Unsöld, A. 1931, *Zeitschrift für Astrophysik*, **1**, 138 and **2**, 209.

Unsöld, A. 1969, *The New Cosmos*, Springer Verlag, Berlin, p. 132.

Unsöld, A. 1948, *Astrophysik*, **25**, 11.

VandenBerg, D. A. and Bell, R. 1985, *Astrophys. J. Suppl.*, **58**, 561.

Weidemann, V. 1975, in *Problems of Stellar Atmospheres and Envelopes*, B. Baschek, W. H. Kegel and G. Traving (eds.), p. 1973, Springer Verlag, New York, Heidelberg, Berlin.

Zhevakin, S. A. 1959, *Russ. Astrophys. J.*, **36**, 996, translated in *Soviet Astrophys. J.*, **3**, 913 (1960).

Other books and textbooks on stellar structure and evolution

Aller, L. and McLaughlin, D. (eds.), *Stellar Structure*, vol. VIII of *Stars and Stellar Systems*. Gen. ed. G. Kuiper and B. Middlehurst, University of Chicago Press, Chicago, 1965.

Bower, R. and Deming, T., *Astrophysics: I. Stars*. Jones and Bartlett Publishers Inc., Boston 1984.

Chandrasekhar, S., *An Introduction to the Study of Stellar Structure*. University of Chicago Press, Chicago, 1939. (Also Dover Publication.)

Chiu, H.-Y. and Muriel, A. (eds.), *Stellar Evolution*. MIT Press (1972).

Clayton, D., *Principles of Stellar Evolution and Nucleosynthesis*. McGraw-Hill, New York, 1968, University of Chicago Press, Chicago, 1983.

Cox, J. and Giuli, R., *Stellar Structure*, vol. I. Physical Principles, Vol. II, Applications to Stars. Gordon and Breach, New York, 1968.

Eddington, A., *The Internal Constitution of Stars*, 1st edition 1926. Dover Publication Inc., New York, 1959.

Kippenhahn, R., *100 Billion Suns*. Basic Books Inc. Publishers, New York, 1983.

Kippenhahn, R. and Weigert, A., *Stellar Structure and Evolution*, Springer-Verlag, Berlin, Heidelberg, New York, 1990.

Motz, L., *Astrophysics and Stellar Structure*. Ginn and Company, Waltham, Massachusetts, 1970.

Novotny, E., *Introduction to Stellar Atmospheres and Interiors*. Oxford University Press, New York, 1973.

Schwarzschild, M., *Structure and Evolution of the Stars*. Princeton University Press, Princeton, New Jersey, 1958 (Dover Publications).

Index